STUDENT SOLUTIONS
VOLUME 1: CHAPTERS 1–16

COLLEGE PHYSICS
A STRATEGIC APPROACH
Second Edition

KNIGHT • JONES • FIELD

LARRY K. SMITH
SNOW COLLEGE

MARLLIN SIMON
AUBURN UNIVERSITY

PAWAN KAHOL
MISSOURI STATE UNIVERSITY

Addison-Wesley

Boston   Columbus   Indianapolis   New York   San Francisco   Upper Saddle River
Amsterdam   Cape Town   Dubai   London   Madrid   Milan   Munich   Paris   Montréal   Toronto
Delhi   Mexico City   São Paulo   Sydney   Hong Kong   Seoul   Singapore   Taipei   Tokyo

Publisher: Jim Smith

Director of Development: Michael Gillespie

Sponsoring Editor: Alice Houston, Ph.D.

Senior Project Editor: Martha Steele

Managing Editor: Corinne Benson

Production Supervisor: Camille Herrera

Production Management and Compositor: GEX Publishing Services

Cover Production: Seventeenth Street Studios

Illustrators: Rolin Graphics

Text and Cover Printer: Edwards Brothers, Inc.

Executive Marketing Manager: Scott Dustan

Cover Photo Credit: Stephen Dalton/Minden Pictures

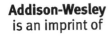

Addison-Wesley
is an imprint of

www.pearsonhighered.com

ISBN 10: 0-321-59629-3

ISBN 13: 978-0-321-59629-1

2 3 4 5 6 7 8 9 10—EB—13 12 11 10

Table of Contents

Preface

This *Student Solutions Manual* is intended to provide you with examples of good problem-solving techniques and strategies. To achieve that, the solutions presented here attempt to:

- Follow, in detail, the problem-solving strategies presented in the text.
- Articulate the reasoning that must be done before computation.
- Illustrate how to use drawings effectively.
- Demonstrate how to utilize graphs, ratios, units, and the many other "tactics" that must be successfully mastered and marshaled if a problem-solving strategy is to be effective.
- Show examples of assessing the reasonableness of a solution.
- Comment on the significance of a solution or on its relationship to other problems.

We recommend you try to solve each problem on your own before you read the solution. Simply reading solutions, without first struggling with the issues, has limited educational value.

As you work through each solution, make sure you understand how and why each step is taken. See if you can understand which aspects of the problem made this solution strategy appropriate. You will be successful on exams not by memorizing solutions to particular problems but by coming to recognize which kinds of problem-solving strategies go with which types of problems.

We have made every effort to be accurate and correct in these solutions. However, if you do find errors or ambiguities, we would be very grateful to hear from you.

REPRESENTING MOTION

Q1.3. Reason:

Assess: The dots are equally spaced until the brakes are applied to the car. Equidistant dots indicate constant average speed. On braking, the dots get closer as the average speed decreases.

Q1.5. Reason: Position refers to the location of an object at a given time relative to a coordinate system. Displacement, on the other hand, is the difference between the object's final position at time t_f and the initial position at time t_i. Displacement is a vector, whose direction is from the initial position toward the final position. An airplane at rest relative to a runway lamp, serving as the origin of our coordinate system, will have a position, called the initial position. The location of the airplane as it takes off may be labeled as the final position. The difference between the two positions, final minus initial, is displacement.

Assess: Some physics texts are not as explicit or clear about this terminology, but it pays off to have clear definitions for terms and to use them consistently.

Q1.9. Reason: Yes, the velocity of an object can be positive during a time interval in which its position is always negative, such as when (in a usual coordinate system with positive to the right) an object is left of the origin, but moving to the right. For example, $x_i = -6.0$ m and $x_f = -2.0$ m. (The magnitude of Δt here is unimportant as long as time goes forward.)

However, the velocity (a vector) is defined to be the displacement (a vector) divided by the time interval (a scalar), and so the velocity *must* have the same sign as the displacement (as long as Δt is positive, which it is when time goes forward). So the answer to the second question is no.

Assess: We see again the importance of defining terms carefully and using them consistently. Students often use physics language incorrectly and then protest, "but you knew what I meant." However, incorrect word usage generally exposes incomplete or incorrect understanding.

Also note that unless stated otherwise, we assume that our coordinate system has positive to the right and that time goes forward (so that Δt is always positive).

Q1.11. Reason:

Assess: The dots get farther apart and the velocity arrows get longer as she speeds up.

Q1.15. Reason:

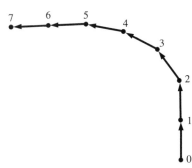

Assess: The car (particle) moves at a constant speed v so the distance between the dots is constant. While turning v remains constant, but the direction of $\vec{v}$ changes.

Q1.19. Reason: Because density is defined to be the ratio of two scalars (mass and volume), it too must be a scalar. **Assess:** This is always true. The ratio of scalars is a scalar. On the other hand, a vector divided by a scalar (as in the definition of velocity) is a vector.

Problems

P1.1. Prepare: Frames of the video are taken at equal intervals of time. As a result, we have a record of the position of the car a successive time equal intervals – this information allows us to construct a motion diagram. **Solve:**

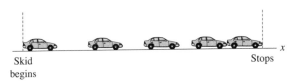

Assess: Once the brakes are applied, the car slows down and travels a smaller distance during each successive time interval until it stops. This is what the car in the figure is doing.

P1.5. Prepare: Displacement is the difference between a final position x_f and an initial position x_i. This can be written as $\Delta x = x_f - x_i$, and we are given that $x_i = 23$ m and that $\Delta x = -45$m. **Solve:** $\Delta x = x_f - x_i$

Since we want to know the final position we solve this for x_f.

$$x_f = x_i + \Delta x$$
$$= 23 \text{ m} + (-45 \text{ m})$$
$$= -22 \text{ m}$$

Assess: A negative displacement means a movement to the left, and Keira has moved left from $x = 23$ m to $x = -22$ m.

P1.7. Prepare: We have been given three different displacements. The problem is straightforward since all the displacements are along a straight east-west line. All we have to do is add the displacements and see where we end up.

Solution: The first displacement is $\Delta \vec{x}_1 = 500$ m east, the second is $\Delta \vec{x}_2 = 400$ m west and the third displacement is $\Delta \vec{x}_3 = 700$ m east. These three displacements are added in the figure below.

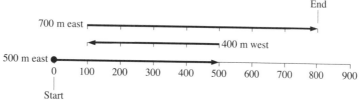

From the figure, note that the result of the sum of the three displacements puts the bee 800 m east of its starting point.
Assess: Knowing what a displacement is and how to add displacements, we are able to obtain the final position of the bee. Since the bee moved 1200 m to the east and 400 m to the west, it is reasonable that it would end up 800 m to the east of the starting point.

P1.9. Prepare: We are asked to rank order three different speeds, so we simply compute each one according to Equation 1.1:

$$\text{speed} = \frac{\text{distance traveled in a given time interval}}{\text{time interval}}$$

Solve: (i) Toy

$$\frac{0.15\text{m}}{2.5\text{s}} = 0.060 \text{ m/s}$$

(ii) Ball

$$\frac{2.3\text{m}}{0.55\text{s}} = 4.2 \text{ m/s}$$

(iii) Bicycle

$$\frac{0.60\text{m}}{0.075\text{s}} = 8.0 \text{ m/s}$$

(iv) Cat

$$\frac{8.0\text{m}}{2.0\text{s}} = 4.0 \text{ m/s}$$

So the order from fastest to slowest is bicycle, ball, cat, and toy.
Assess: We reported all answers to two significant figures as we should according to the significant figure rules. The result is probably what we would have guessed before solving the problem, although the cat and ball are close. These numbers all seem reasonable for the respective objects.

P1.13. Prepare: In this problem we are given $x_i = 2.1$ m and $x_f = 7.3$ m as well as $v = 0.35$ m/s and asked to solve for Δt.
Solve: We first solve for Δt in $v = \Delta x/\Delta t$ and then apply $\Delta x = x_f - x_i$:

$$\Delta t = \frac{\Delta x}{v} = \frac{x_f - x_i}{v} = \frac{7.3 \text{ m} - 2.1 \text{ m}}{0.35 \text{ m/s}} = \frac{5.2 \text{ m}}{0.35 \text{ m/s}} = 15 \text{ s}$$

Assess: 15 s seems like a long time for a ball to roll, but it is going fairly slowly, so the answer is reasonable.

P1.15. Prepare: We first collect the necessary conversion factors: 1 in = 2.54 cm; 1 cm = 10^{-2} m; 1 ft = 12 in; 39.37 in = 1 m; 1 mi = 1.609 km; 1 km = 10^3 m; 1 h = 3600 s.

Solve:

$$\text{(a)} \quad 8.0 \text{ in} = 8.0 \text{ (in)} \left(\frac{2.54 \text{ cm}}{1 \text{ in}} \right) \left(\frac{10^{-2} \text{ m}}{1 \text{ cm}} \right) = 0.20 \text{ m}$$

$$\text{(b)} \quad 66 \text{ ft/s} = 66 \left(\frac{\text{ft}}{\text{s}} \right) \left(\frac{12 \text{ in}}{1 \text{ ft}} \right) \left(\frac{1 \text{ m}}{39.37 \text{ in}} \right) = 20 \text{ m/s}$$

$$\text{(c)} \quad 60 \text{ mph} = 60 \left(\frac{\text{mi}}{\text{h}} \right) \left(\frac{1.609 \text{ km}}{1 \text{ mi}} \right) \left(\frac{10^3 \text{ m}}{1 \text{ km}} \right) \left(\frac{1 \text{ h}}{3600 \text{ s}} \right) = 27 \text{ m/s}$$

P1.19. Prepare: Review the rules for significant figures in Section 1.4 of the text, paying particular attention to any zeros and whether or not that they are significant.
Solve: (a) The number 0.621 has three significant figures.
(b) The number 0.006200 has four significant figures.
(c) The number 1.0621 has five significant figures.
(d) The number 6.21×10^3 has three significant figures.
Assess: In part (b), The initial two zeroes place the decimal point. The last two zeroes do not have to be there, but when they are they are significant.

P1.21. Prepare: Table 1.3 supplies the conversion factor we need: 1 ft = 0.305 m.
Solve:

$$1250 \text{ ft} = 1250 \text{ ft} \left(\frac{0.305 \text{ m}}{1 \text{ ft}} \right) = 381 \text{ m} = 3.81 \times 10^2 \text{ m}$$

Assess: Field and track fans will quickly recognize that this answer is reasonable. They will note that 1250 ft is a little less than a quarter of a mile and 381 m is a little less than 440 m which is a little more than once around a quarter mile track.

P1.25. Prepare: Think about how often you cut your fingernails, and how much you cut off each time. Depending on how even and trim you like to keep them you might clip them every 10 days or so; you might clip off about 1 mm every 10 days.
Solve: The speed $v_x = \Delta x / \Delta t$ of the tip of your fingernails relative to your finger is then about 1 mm/10 d.

$$v_x = \frac{1 \text{ mm}}{10 \text{ d}} \left(\frac{1 \text{ m}}{1000 \text{ mm}} \right) \left(\frac{1 \text{ d}}{24 \text{ h}} \right) \left(\frac{1 \text{ h}}{60 \text{ min}} \right) \left(\frac{1 \text{ min}}{60 \text{ s}} \right) \approx 1 \times 10^{-9} \text{ m/s} \qquad v_x = \frac{1 \text{ mm}}{10 \text{ d}} \left(\frac{1000 \text{ } \mu\text{m}}{1 \text{ mm}} \right) \left(\frac{1 \text{ d}}{24 \text{ h}} \right) \approx 4 \text{ } \mu\text{m/h}$$

Assess: There are various factors that affect the growth rate of fingernails, such as age, sex, and season. Fingernails tend to grow faster than toenails, and not all fingernails grow at the same rate. But our answer seems about right and generally agrees with values found in a Web search. At this rate it takes about 6 months to completely replace a fingernail.

P1.27. Prepare: Joe and Max went at right angles to each other, so we can use the Pythagorean theorem to compute the distance between them.
Solve: $d = \sqrt{(3.25 \text{ km})^2 + (0.55 \text{ km})^2} = 3.30 \text{ km}$
Assess: It is right that the difference between the two is greater than the distance either of them went; the answer seems about right for a skinny triangle.

P1.29. Prepare: The displacement is the hypotenuse of a right triangle whose legs are 6 m (half the width of the garden) and 8 m, directed from x_i at the top of the tree to x_f at the top of the flower.

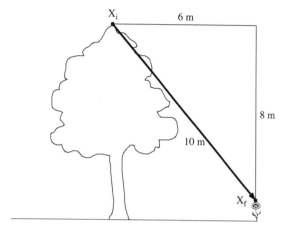

Solve: The length of the hypotenuse is $\sqrt{(6 \text{ m})^2 + (8 \text{ m})^2} = 10$ m.

Assess: Displacement is a vector and does have direction (as mentioned in the prepare step), but we were only asked for the magnitude of the displacement.

P1.31. Prepare: Knowing that the total trip consists of two displacements, we can add the two displacements to determine the total displacement and hence the distance of the goose from its original position. A quick sketch will help you visualize the two displacements and the total displacement.

Solution: The distance of the goose from its original position is the magnitude of the total displacement vector. This is determined as follows:

$$d = \sqrt{(32 \text{ km})^2 + (20 \text{ km})^2} = 38 \text{ km}$$

Assess: A quick look at your sketch shows that the total distance should be larger than the largest leg of the trip and this is the case.

P1.35. Prepare: The watermelon, represented as a particle, falls freely and speeds up during its downward motion along the y-direction. The average velocity vectors are thus of increasing length, giving an acceleration vector that is pointed in the downward direction. The following diagram shows the watermelon's motion including a pictorial representation and a list of values.

Solve:

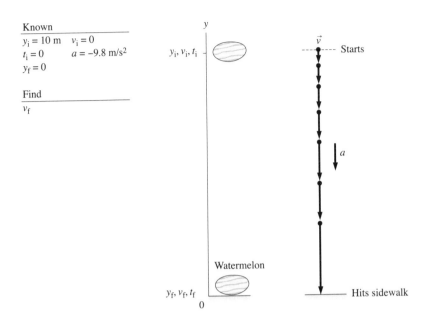

Assess: As a result of the acceleration due to gravity, as the watermelon falls, its velocity and the distance between the position dots increases. This is the case in the figure show.

P1.39. Prepare: Represent the ball as a particle which is moving along the ramp defined as the *x*-axis. As the ball rolls up the ramp, it slows down indicating accelerating down the ramp. The direction of the acceleration vector at the point where the slope changes is due to the fact that the average velocity vector along the ramp is smaller than the vector along the floor.

Solve:

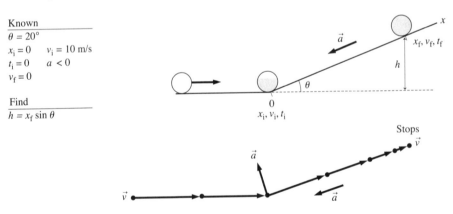

Assess: While the ball is traveling across the horizontal section, the velocity vector and the spacing of the position dots is constant. As the ball makes the transition from the horizontal section to the ramp, there is an acceleration in the direction of the change in velocity. As the ball travels up the ramp the negative acceleration causes the velocity vectors to decrease in length and the position dots to get closer together.

P1.41. Prepare: Knowing the dots represent the position of an object at equal time intervals and the vectors represent the velocity of the object at these times, we can construct a situation to match the motion diagram.
Solve: Rahul was coasting on interstate highway I-44 from Tulsa to Springfield at 70 mph. Seeing an accident at a distance of 200 feet in front of him, he began to brake. What steady deceleration will bring him to a stop at the accident site?
Assess: Since the position dots are initially equally spaced and the first few velocity vectors have the same length, this is consistent with Rahul initially traveling at a constant velocity. The fact that the dots get closer

together and the velocity vectors get shorter is consistent with Rahul's braking. The fact that there is no velocity vector associated with the last dot is consistent with the fact that he braked to a stop.

P1.47.   Prepare: Given the speed of light (3.0×10^8 m/s) and a time of (1.0 ns), we are asked to compute the distance light can travel in the given time. To solve the problem, two preliminary unit conversions will be helpful: $1.0 \text{ ns} = 1.0 \times 10^{-9}$ s and $1 \text{m} = 39.37$ in

Solve:

$$\text{speed} = \frac{\text{distance traveled in a given time interval}}{\text{time interval}}$$

Solving for distance, obtain

$$\text{distance} = (\text{speed})(\text{time}) = \left(3.0 \times 10^8 \, \frac{\text{m}}{\text{s}} \right) (1.0 \text{ ns}) \left(1.0 \times 10^{-9} \, \frac{\text{s}}{\text{ns}} \right) \left(39.37 \frac{\text{in}}{\text{m}} \right) = 12 \text{ in}$$

Assess: Notice that nanoseconds, seconds, and meters cancel out neatly, leaving the desired distance (in). Think about the answer: In a billionth of a second light can go 1 foot; this is a useful tidbit to tuck into your brain. Just as we express large astronomical distance in terms of "light years" (a unit of distance—not time; the distance light travels in a year), we can now somewhat whimsically refer to a foot as a "light nanosecond."

P1.51.   Prepare: Knowing that speed is distance divided by time, the distance is the circumference of a circle of radius 93,000,000 miles and the time is one year, we can determine the speed of the earth orbiting the sun. In order to get an answer in m/s, some unit conversion will be required.

Solve: The speed of the earth in its orbit about the sun may be determined by

$$v = \frac{\text{distance}}{\text{time}} = \frac{2\pi r}{t} = \frac{2\pi \left(9.3 \times 10^7 \text{mi} \right)}{1 \text{ yr}} \left(\frac{1 \text{ yr}}{3.16 \times 10^7 \text{s}} \right) \left(\frac{1.61 \times 10^3 \text{ m}}{1 \text{ mi}} \right) = 3.0 \times 10^4 \text{m/s}$$

Assess: All of the unit conversions are correct, after units are canceled we obtain the desired units (m/s) and we are expecting a large number.

P1.55.   Prepare: Knowing that a motion diagram consists of dots that represent the position of the object at equal intervals of time and velocity vectors that represent the velocity at each position, we can construct a motion diagram for this situation.

Solve: **(a)** The figure below shows the motion diagram for the nerve impulses traveling along fibers A, B, and C.

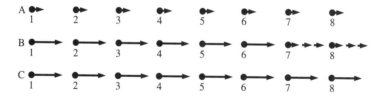

The figure shows the position of each axon and the speed of the nerve impulse as it travels through that axon. In fiber A, the impulses travel slow (2.0 m/s) through all eight axons. In fiber B, the nerve impulses travel fast (25 m/s) through the first six axons and then slow (2.0 m/s) through the last two. In fiber C, the impulses travel fast (25 m/s) through all the axons. For case A, the time interval used is the time for the impulse to travel the length of one axon. For case B, some liberties have been taken. Initially, the time interval is the time for the impulse to travel the length of one axon. However, if this was continued to scale, for the last two axons we would need over 10 dots and velocity vectors in the length of each axon. As a result, for the last two axons the dots are closer together and the velocity vectors are shorter, but they are not to scale.

(b)   Nerve impulses in the fully myelinated fiber will get to the right end first.

(c)   Nerve impulses in the unmyelinated fiber will get to the right end last.

Assess: The impulses travel through fiber A at a slow constant speed. As a result the position dots should be uniformly spaced and the velocity vectors should all be the same length. The impulses travel through fiber C at a fast constant speed. As a result the position dots should be uniformly spaced and the velocity vectors should all be the same length. For fiber B, the impulses travel fast for six axons and then slow for two axons. As a result the position dots should be uniformly spaced for the first six axons and then uniformly spaced (but closer together) for the last two axons. In addition the velocity vectors should be constant length for the first six axons and constant in length (but shorter) for the last two axons.

P1.59.   Prepare: The best way to prepare for this problem is to draw a diagram. In the triangle, the height h of the tree that we want to know is the side opposite to the given angle.

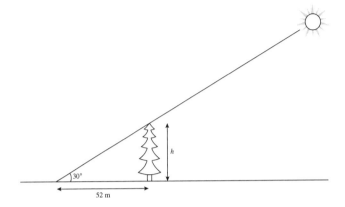

Solve:

$$\tan\theta = \frac{\text{opposite}}{\text{adjacent}}$$

$$\text{opposite} = \text{adjacent} \times \tan\theta$$

$$h = 52 \text{ m} \times \tan 30° = 30 \text{ m}$$

Assess: The shadow is longer than the height of the tree because the sun is low in the sky.

Q2.1. Reason: The elevator must speed up from rest to cruising velocity. In the middle will be a period of constant velocity, and at the end a period of slowing to a rest.

The graph must match this description. The value of the velocity is zero at the beginning, then it increases, then, during the time interval when the velocity is constant, the graph will be a horizontal line. Near the end the graph will decrease and end at zero.

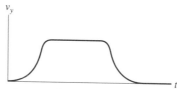

Assess: After drawing velocity-versus-time graphs (as well as others), stop and think if it matches the physical situation, especially by checking end points, maximum values, places where the slope is zero, etc. This one passes those tests.

Q2.5. Reason: Yes. The acceleration vector will point south when the car is slowing down while traveling north.

Assess: The acceleration vector will always point in the direction opposite the velocity vector in straight line motion if the object is slowing down. Feeling good about this concept requires letting go of the common every day (mis)usage where velocity and acceleration are sometimes treated like synonyms. Physics definitions of these terms are more precise and when discussing physics we need to use them precisely.

Q2.7. Reason: We will neglect air resistance, and thus assume that the ball is in free fall.

(a) $-g$ After leaving your hand the ball is traveling up but slowing, therefore the acceleration is down (*i.e.*, negative).

(b) $-g$ At the very top the velocity is zero, but it had previously been directed up and will consequently be directed down, so it is changing direction (*i.e.*, accelerating) down.

(c) $-g$ Just before hitting the ground it is going down (velocity is down) and getting faster; this also constitutes an acceleration down.

Assess: As simple as this question is, it is sure to illuminate a student's understanding of the difference between velocity and acceleration. Students would be wise to dwell on this question until it makes complete sense.

Q2.13. Reason: (a) For the velocity to be constant, the velocity-versus-time graph must have zero slope. Looking at the graph, there are three time intervals where the graph has zero slope: segment A, segment D and segment F.

(b) For an object to be speeding up, the magnitude of the velocity of the object must be increasing. When the slope of the lines on the graph is nonzero, the object is accelerating and therefore changing speed.

Consider segment B. The velocity is positive while the slope of the line is negative. Since the velocity and acceleration are in opposite directions, the object is slowing down. At the start of segment B, we can see the velocity is +2 m/s, while at the end of segment B the velocity is 0 m/s.

During segment E the slope of the line is positive which indicates positive acceleration, but the velocity is negative. Since the acceleration and velocity are in opposite directions, the object is slowing here also. Looking at the graph at the beginning of segment E the velocity is –2 m/s, which has a magnitude of 2 m/s. At the end of segment E the velocity is 0 m/s, so the object has slowed down.

Consider segment C. Here the slope of the line is negative and the velocity is negative. The velocity and acceleration are in the same direction so the object is speeding up. The object is gaining velocity in the negative

direction. At the beginning of that segment the velocity is 0 m/s, and at the end the velocity is –2 m/s, which has a magnitude of 2 m/s.
(c) In the analysis for part **(b)**, we found that the object is slowing down during segments B and E.
(d) An object standing still has zero velocity. The only time this is true on the graph is during segment F, where the line has zero slope, and is along $v = 0$ m/s.
(e) For an object to moving to the right, the convention is that the velocity is positive. In terms of the graph, positive values of velocity are above the time axis. The velocity is positive for segments A and B. The velocity must also be greater than zero. Segment F represents a velocity of 0 m/s.
Assess: Speed is the magnitude of the velocity vector. Compare to Conceptual Example 2.6 and also Question 2.2.

Problems

P2.1. Prepare: The car is traveling to the left toward the origin, so its position decreases with increase in time.
Solve: (a)

Time t (s)	Position x (m)
0	1200
1	975
2	825
3	750
4	700
5	650
6	600
7	500
8	300
9	0

(b)

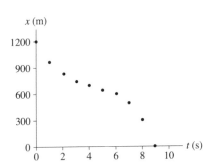

Assess: A car's motion traveling down a street can be represented at least three ways: a motion diagram, position-versus-time data presented in a table (part **(a)**), and a position-versus-time graph (part **(b)**).

P2.3. Prepare: The graph represents an object's motion along a straight line. The object is in motion for the first 300 s and the last 200 s, and it is not moving from $t = 300$ s to $t = 400$ s.
Solve: A forgetful physics professor goes for a walk on a straight country road. Walking at a constant speed, he covers a distance of 300 m in 300 s. He then stops and watches the sunset for 100 s. Realizing that it is getting dark, he walks faster back to his house covering the same distance in 200 s.
Assess: The slope of the graph is positive up to $t = 300$ s, so the velocity is positive and motion is to the right. However, the slope is negative from $t = 400$ s to $t = 600$ s, so the velocity is negative and motion is to the left. Furthermore, because slope for the latter time interval is more than the former, motion to the left is faster than motion to the right.

P2.7. Prepare: To get a position from a velocity graph we count the area under the curve.
Solve:
(a)

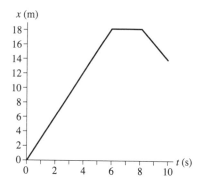

(b) We need to count the area under the velocity graph (area below the *x*-axis is subtracted). There are 18 m of area above the axis and 4 m of area below. $18\,\text{m} - 4\,\text{m} = 14\,\text{m}$.

Assess: These numbers seem reasonable; a mail carrier could back up 4 m. It is also important that the problem state what the position is at $t = 0$, or we wouldn't know how high to draw the position graph.

P2.11. Prepare: A visual overview of Alan's and Beth's motion that includes a pictorial representation, a motion diagram, and a list of values is shown below. Our strategy is to calculate and compare Alan's and Beth's time of travel from Los Angeles to San Francisco.

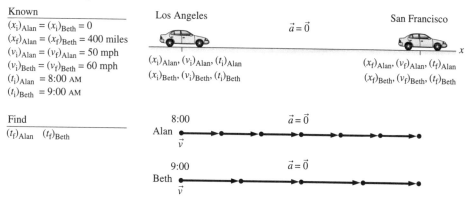

Solve: Beth and Alan are moving at a constant speed, so we can calculate the time of arrival as follows:

$$v = \frac{\Delta x}{\Delta t} = \frac{x_f - x_i}{t_f - t_i} \Rightarrow t_f = t_i + \frac{x_f - x_i}{v}$$

Using the known values identified in the pictorial representation, we find

$$\left(t_f\right)_{\text{Alan}} = \left(t_i\right)_{\text{Alan}} + \frac{\left(x_f\right)_{\text{Alan}} - \left(x_i\right)_{\text{Alan}}}{v} = 8{:}00 \text{ AM} + \frac{400 \text{ mile}}{50 \text{ miles/hour}} = 8{:}00 \text{ AM} + 8 \text{ hr} = 4{:}00 \text{ PM}$$

$$\left(t_f\right)_{\text{Beth}} = \left(t_i\right)_{\text{Beth}} + \frac{\left(x_f\right)_{\text{Beth}} - \left(x_i\right)_{\text{Beth}}}{v} = 9{:}00 \text{ AM} + \frac{400 \text{ mile}}{60 \text{ miles/hour}} = 9{:}00 \text{ AM} + 6.67 \text{ hr} = 3{:}40 \text{ PM}$$

(a) Beth arrives first.
(b) Beth has to wait 20 minutes for Alan.
Assess: Times of the order of 7 or 8 hours are reasonable in the present problem.

P2.15. Prepare: Assume v_x is constant so the ratio $\dfrac{\Delta x}{\Delta t}$ is also constant.

Solve:

(a)

$$\frac{30 \text{ m}}{3.0 \text{ s}} = \frac{\Delta x}{1.5 \text{ s}} \quad \Rightarrow \quad \Delta x = 1.5 \text{ s}\left(\frac{30 \text{ m}}{3.0 \text{ s}}\right) = 15 \text{ m}$$

(b)

$$\frac{30 \text{ m}}{3.0 \text{ s}} = \frac{\Delta x}{9.0 \text{ s}} \quad \Rightarrow \quad \Delta x = 9.0 \text{ s}\left(\frac{30 \text{ m}}{3.0 \text{ s}}\right) = 90 \text{ m}$$

Assess: Setting up the ratio allows us to easily solve for the distance traveled in any given time.

P2.17. Prepare: The graph in Figure P2.17 shows distinct slopes in the time intervals: 0 – 1s, 1 s – 2 s, and 2 s – 4 s. We can thus obtain the velocity values from this graph using $v = \Delta x/\Delta t$.

Solve: (a)

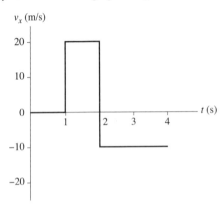

(b) There is only one turning point. At $t = 2$ s the velocity changes from +20 m/s to −10 m/s, thus reversing the direction of motion. At $t = 1$ s, there is an abrupt change in motion from rest to +20 m/s, but there is no reversal in motion.

Assess: As shown above in **(a)**, a positive slope must give a positive velocity and a negative slope must yield a negative velocity.

P2.21. Prepare: Please refer to Figure P2.21. The graph in Figure P2.21 shows distinct slopes in the time intervals: 0 – 2 s and 2 s – 4 s. We can thus obtain the acceleration values from this graph using $a_x = \Delta v_x/\Delta t$. A linear decrease in velocity from $t = 0$ s to $t = 2$ s implies a constant negative acceleration. On the other hand, a constant velocity between $t = 2$ s and $t = 4$ s means zero acceleration.

Solve:

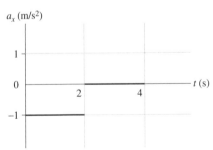

P2.25. Prepare: From a velocity-versus-time graph we find the acceleration by computing the slope. We will compute the slope of each straight-line segment in the graph.

$$a_x = \frac{(v_x)_f - (v_x)_i}{t_f - t_i}$$

The trickiest part is reading the values off of the graph.
Solve: (a)

$$a_x = \frac{5.5 \text{ m/s} - 0.0 \text{ m/s}}{0.9 \text{ s} - 0.0 \text{ s}} = 6.1 \text{ m/s}^2$$

(b)

$$a_x = \frac{9.3 \text{ m/s} - 5.5 \text{ m/s}}{2.4 \text{ s} - 0.9 \text{ s}} = 2.5 \text{ m/s}^2$$

(c)

$$a_x = \frac{10.9 \text{ m/s} - 9.3 \text{ m/s}}{3.5 \text{ s} - 2.4 \text{ s}} = 1.5 \text{ m/s}^2$$

Assess: This graph is difficult to read to more than one significant figure. I did my best to read a second significant figure but there is some estimation in the second significant figure.
It takes Carl Lewis almost 10 s to run 100 m, so this graph covers only the first third of the race. Were the graph to continue, the slope would continue to decrease until the slope is zero as he reaches his (fastest) cruising speed. Also, if the graph were continued out to the end of the race, the area under the curve should total 100 m.

P2.27. Prepare: Acceleration is the rate of change of velocity.

$$a_x = \frac{\Delta v_x}{\Delta t}$$

Where $\Delta v_x = 4.0$ m/s and $\Delta t = 0.11$ s.

We will then use that acceleration in Equation 2.14 (a special case of Equation 2.12) to compute the displacement during the strike:

$$\Delta x = \frac{1}{2} a_x (\Delta t)^2$$

where we are justified in using the special case because $(v_x)_i = 0.0$ m/s.
Solve: (a)

$$a_x = \frac{\Delta v_x}{\Delta t} = \frac{4.0 \text{ m/s}}{0.11 \text{ s}} = 36 \text{ m/s}^2$$

(b)

$$\Delta x = \frac{1}{2} a_x (\Delta t)^2 = \frac{1}{2} (36 \text{ m/s}^2)(0.11 \text{ s})^2 = 0.22 \text{ m}$$

Assess: The answer is remarkable but reasonable. The pike strikes quickly and so is able to move 0.22 m in 0.11 s, even starting from rest. The seconds squared cancel in the last equation.

P2.31. Prepare: The kinematic equation that relates velocity, acceleration, and distance is $(v_x)_f^2 = (v_x)_i^2 + 2a_x \Delta x$. Solve for Δx.

$$\Delta x = \frac{(v_x)_f^2 - (v_x)_i^2}{2a_x}$$

Note that $(v_x)_i^2 = 0$ for both planes.

Solve: The accelerations are same, so they cancel.

$$\frac{\Delta x_{\text{jet}}}{\Delta x_{\text{prop}}} = \frac{\left(\dfrac{(v_x)_f^2}{2a_x}\right)_{\text{jet}}}{\left(\dfrac{(v_x)_f^2}{2a_x}\right)_{\text{prop}}} = \frac{\left((v_x)_f\right)^2_{\text{jet}}}{\left((v_x)_f\right)^2_{\text{prop}}} = \frac{\left((2v_x)_f\right)^2_{\text{prop}}}{\left((v_x)_f\right)^2_{\text{prop}}} = 4 \quad \Rightarrow \quad \Delta x_{\text{jet}} = 4\Delta x_{\text{prop}} = 4(1/4 \text{ mi}) = 1 \text{ mi}$$

Assess: It seems reasonable to need a mile for a passenger jet to take off.

P2.33. Prepare: A visual overview of the car's motion that includes a pictorial representation, a motion diagram, and a list of values is shown below. We label the car's motion along the *x*-axis. For the driver's maximum (constant) deceleration, kinematic equations are applicable. This is a two-part problem. We will first find the car's displacement during the driver's reaction time when the car's deceleration is zero. Then we will find the displacement as the car is brought to rest with maximum deceleration.

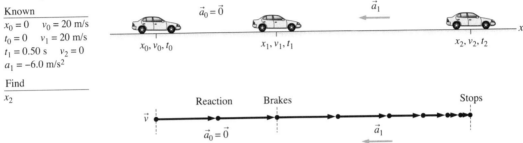

Known
$x_0 = 0$ $v_0 = 20$ m/s
$t_0 = 0$ $v_1 = 20$ m/s
$t_1 = 0.50$ s $v_2 = 0$
$a_1 = -6.0$ m/s^2

Find
x_2

Solve: During the reaction time when $a_0 = 0$, we can use

$$x_1 = x_0 + v_0(t_1 - t_0) + \frac{1}{2}a_0(t_1 - t_0)^2$$
$$= 0 \text{ m} + (20 \text{ m/s})(0.50 \text{ s} - 0 \text{ s}) + 0 \text{ m} = 10 \text{ m}$$

During deceleration,

$$v_2^2 = v_1^2 + 2a_1(x_2 - x_1) \qquad 0 = (20 \text{ m/s})^2 + 2(-6.0 \text{ m/s}^2)(x_2 - 10 \text{ m}) \Rightarrow x_2 = 43 \text{ m}$$

She has 50 m to stop, so she can stop in time.
Assess: While driving at 20 m/s or 45 mph, a reaction time of 0.5 s corresponds to a distance of 33 feet or only two lengths of a typical car. Keep a safe distance while driving!

P2.37. Prepare: We will use the equation for constant acceleration to find out how far the sprinter travels during the acceleration phase. Use Equation 2.11 to find the acceleration.

$$v_x = a_x t_1 \qquad \text{where } v_0 = 0 \text{ and } t_0 = 0$$

$$a_x = \frac{v_x}{t_1} = \frac{11.2 \text{ m/s}}{2.14 \text{ s}} = 5.23 \text{ m/s}^2$$

Solve: The distance traveled during the acceleration phase will be

$$\Delta x = \frac{1}{2}a_x(\Delta t)^2$$
$$= \frac{1}{2}(5.23 \text{ m/s}^2)(2.14 \text{ s})^2$$
$$= 12.0 \text{ m}$$

The distance left to go at constant velocity is $100 \text{ m} - 12.0 \text{ m} = 88.0 \text{ m}$. The time this takes at the top speed of 11.2 m/s is

$$\Delta t = \frac{\Delta x}{v_x} = \frac{88.0 \text{ m}}{11.2 \text{ m/s}} = 7.86 \text{ s}$$

The total time is $2.14 \text{ s} + 7.86 \text{ s} = 10.0 \text{ s}$.

Assess: This is indeed about the time it takes a world-class sprinter to run 100 m (the world record is a bit under 9.8 s).

Compare the answer to this problem with the accelerations given in Problem 2.25 for Carl Lewis.

P2.39. Prepare: Review the related "Try It Yourself" in the chapter.

We will assume that, as stated in the chapter, the bill is held at the top, and the other person's fingers are bracketing the bill at the bottom.

Call the initial position of the top of the bill the origin, $y_o = 0.0 \text{ m}$, and, for convenience, call the down direction positive.

In free fall the acceleration a_y will be 9.8 m/s^2.

The length of the bill will be Δy, the distance the top of the bill can fall from rest in 0.25 s.
Use Equation 2.14.
Solve:

$$\Delta y = \frac{1}{2} a_y (\Delta t)^2 = \frac{1}{2}(9.8 \text{ m/s}^2)(0.25 \text{ s})^2 = 0.31 \text{ m}$$

Assess: This is about twice as long as real bills are (they are really 15 cm long), so if a typical reaction time is 0.25 s, then almost no one would catch one in this manner. To catch a bill as small as real bills, one would need a reaction time of 0.13 s.

P2.41. Prepare: If we ignore air resistance then the only force acting on both balls after they leave the hand (before they land) is gravity; they are therefore in free fall.

Think about ball A's velocity. It decreases until it reaches the top of its trajectory and then increases in the downward direction as it descends. When it gets back to the level of the student's hand it will have the same speed downward that it had initially going upward; it is therefore now just like ball B (only later).

Solve: (a) Because both balls are in free fall they must have the same acceleration, both magnitude and direction, 9.8 m/s^2, down.

(b) Because ball B has the same downward speed when it gets back to the level of the student that ball A had, they will have the same speed when they hit the ground.

Assess: Draw a picture of ball B's trajectory and draw velocity vector arrows at various points of its path.

Air resistance would complicate this problem significantly.

P2.45. Prepare: There are several steps in this problem, so first draw a picture and, like the examples in the book, list the known quantities and what we need to find.

Call the pool of water the origin and call $t = 0$ s when the first stone is released. We will assume both stones are in free fall after they leave the climber's hand, so $a_y = -g$. Let a subscript 1 refer to the first stone and a 2 refer to the second.

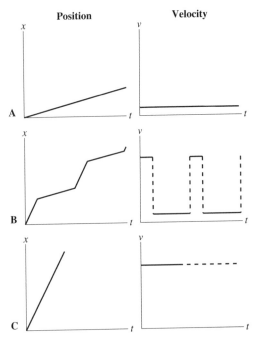

Known
$(y_1)_i = 50$ m
$(y_2)_i = 50$ m
$(y_1)_f = 0.0$ m
$(y_2)_f = 0.0$ m
$(y_1)_i = -2.0$ m/s
$(t_2)_f = (t_1)_f$; simply call this t_f
$(t_2)_i = 1.0$ s

Find
$(t_2)_f$ or t_f
$(v_2)_i$
$(v_1)_f$
$(v_2)_f$

Solve: **(a)** Using $(t_1)_i = 0$

$$(y_1)_f = (y_1)_i + (v_1)_i \Delta t + \frac{1}{2} a_y \Delta t^2$$

$$0.0 \text{ m} = 50 \text{ m} + (-2 \text{ m/s})t_f + \frac{1}{2}(-g)t_f^2$$

$$0.0 \text{ m} = 50 \text{ m} - (2 \text{ m/s})t_f - (4.9 \text{ m/s}^2)t_f^2$$

Solving this quadratic equation gives two values for t_f: 3.0 s and -3.4 s, the second of which (being negative) is outside the scope of this problem.
Both stones hit the water at the same time, and it is at $t = 3.0$ s, or 3.0 s after the first stone is released.

(b) For the second stone $\Delta t_2 = t_f - (t_2)_i = 3.0 \text{ s} - 1.0 \text{ s} = 2.0 \text{ s}$. We solve now for $(v_2)_i$.

$$(y_2)_f = (y_2)_i + (v_2)_i \Delta t + \frac{1}{2} a_y \Delta t^2$$

$$0.0 \text{ m} = 50 \text{ m} + (v_2)_i \Delta t_2 + \frac{1}{2}(-g)\Delta t_2^2$$

$$0.0 \text{ m} = 50 \text{ m} + (v_2)_i (2.0 \text{ s}) - (4.9 \text{ m/s}^2)(2.0 \text{ s})^2$$

$$(v_2)_i = \frac{-50 \text{ m} + (4.9 \text{ m/s}^2)(2.0 \text{ s})^2}{2.0 \text{ s}} = -15.2 \text{ m/s}$$

Thus, the second stone is thrown down at a speed of 15 m/s.
(c) Equation 2.11 allows us to compute the final speeds for each stone.

$$(v_y)_f = (v_y)_i + a_y \Delta t$$

For the first stone (which was in the air for 3.0 s):

$$(v_1)_f = -2.0 \text{ m/s} + (-9.8 \text{ m/s}^2)(3.0 \text{ s}) = -31.4 \text{ m/s} \approx -31 \text{ m/s}$$

The speed is the magnitude of this velocity, or 31 m/s.
For the second stone (which was in the air for 2.0 s):

$$(v_2)_f = -15.2 \text{ m/s} + (-9.8 \text{ m/s}^2)(2.0 \text{ s}) = -34.8 \text{ m/s} \approx -35 \text{ m/s}$$

The speed is the magnitude of this velocity, or 35 m/s.

Assess: The units check out in each of the previous equations. The answers seem reasonable. A stone dropped from rest takes 3.2 s to fall 50 m; this is comparable to the first stone, which was able to fall the 50 m in only 3.0 s because it started with an initial velocity of −2.0 m/s. So we are in the right ballpark. And the second stone would have to be thrown much faster to catch up (because the first stone is accelerating).

P2.47. Prepare: The position graphs of the nerve impulses will be strictly increasing, since the impulse does, in fact, always travel to the right in this context. The graphs will consist of straight-line segments whose slope depends on the speed (i.e., whether the nerve is myelinated or not).

The velocity graphs will be constant horizontal line segments whose height depends on the speed (i.e., whether the nerve is myelinated or not).

Solve:

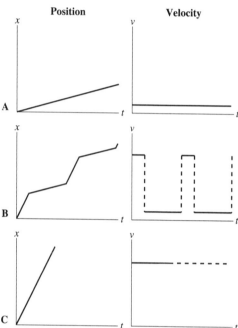

Assess: The velocity graph for fiber C would continue at the same speed if we draw it for a longer fiber than 6 axons.

P2.51. Prepare: We assume that the track, except for the sticky section, is frictionless and aligned along the *x*-axis. Because the motion diagram of Figure P2.51 is made at two frames of film per second, the time interval between consecutive ball positions is 0.5 s.

Solve: (a)

Times (s)	Position
0	−4.0
0.5	−2.0
1.0	0
1.5	1.8
2.0	3.0
2.5	4.0
3.0	5.0
3.5	6.0
4.0	7.0

(b)

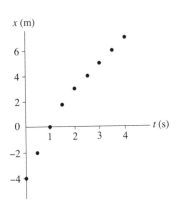

(c) $\Delta x = x$ (at $t = 1$ s) $- x$ (at $t = 0$ s) $= 0$ m $- (-4$ m) $= 4$ m.

(d) $\Delta x = x$ (at $t = 4$ s) $- x$ (at $t = 2$ s) $= 7$ m $- 3$ m $= 4$ m.

(e) From $t = 0$ s to $t = 1$ s, $v_s = \Delta x / \Delta t = 4$ m/s.

(f) From $t = 2$ s to $t = 4$ s, $v_x = \Delta x / \Delta t = 2$ m/s.

(g) The average acceleration is

$$a = \frac{\Delta v}{\Delta t} = \frac{2 \text{ m/s} - 4 \text{ m/s}}{2 \text{ s} - 1 \text{ s}} = -2 \text{ m/s}^2$$

Assess: The sticky section has decreased the ball's speed from 4 m/s, to 2 m/s, which is a reasonable magnitude.

P2.53. Prepare: We will represent the jetliner's motion to be along the *x*-axis.
Solve: (a) To convert 80 m/s to mph, we calculate 80 m/s $\times$ 1 mi/1609 m $\times$ 3600 s/h = 180 mph.
(b) Using $a_x = \Delta v / \Delta t$, we have,

$$a_x (t = 0 \text{ to } t = 10 \text{ s}) = \frac{23 \text{ m/s} - 0 \text{ m/s}}{10 \text{ s} - 0 \text{ s}} = 2.3 \text{ m/s}^2 \qquad a_x (t = 20 \text{ s to } t = 30 \text{ s}) = \frac{69 \text{ m/s} - 46 \text{ m/s}}{30 \text{ s} - 20 \text{ s}} = 2.3 \text{ m/s}^2$$

For all time intervals a_x is 2.3 m/s^2.
(c) Because the jetliner's acceleration is constant, we can use kinematics as follows:

$$(v_x)_f = (v_x)_i + a_x(t_f - t_i) \Rightarrow 80 \text{ m/s} = 0 \text{ m/s} + (2.3 \text{ m/s}^2)(t_f - 0 \text{ s}) \Rightarrow t_f = 34.8 \text{ s} = 35 \text{ s}$$

(d) Using the above values, we calculate the takeoff distance as follows:

$$x_f = x_i + (v_x)_i(t_f - t_i) + \frac{1}{2}a_x(t_f - t_i)^2 = 0 \text{ m} + (0 \text{ m/s})(34.8 \text{ s}) + \frac{1}{2}(2.3 \text{ m/s}^2)(34.8 \text{ s})^2 = 1390 \text{ m}$$

For safety, the runway should be 3 $\times$ 1390 m = 4170 m or 2.6 mi. This is longer than the 2.5 mi long runway, so the takeoff is not safe.

P2.59. Prepare: Fleas are amazing jumpers; they can jump several times their body height — something we cannot do.
We assume constant acceleration so we can use the equations in Table 2.4. The last of the three relates the three variables we are concerned with in part **(a)**: speed, distance (which we know), and acceleration (which we want).

$$(v_y)_f^2 = (v_y)_i^2 + 2a_y \Delta y$$

In part **(b)** we use the first equation in Table 2.4 because it relates the initial and final velocities and the acceleration (which we know) with the time interval (which we want).

$$(v_y)_f = (v_y)_i + a_y \Delta t$$

Part (c) is about the phase of the jump *after* the flea reaches takeoff speed and leaves the ground. So now it is $(v_y)_i$, that is 1.0 m/s instead of $(v_y)_f$. And the acceleration is not the same as in part (a)—it is now $-g$ (with the positive direction up) since we are ignoring air resistance. We do not know the time it takes the flea to reach maximum height, so we employ the last equation in Table 2.4 again because we know everything in that equation except Δy.

Solve: (a) Use $(v_y)_i = 0.0$ m/s and rearrange the last equation in Table 2.4.

$$a_y = \frac{(v_y)_f^2}{2\Delta y} = \frac{(1.0 \text{ m/s})^2}{2(0.50 \text{ mm})}\left(\frac{1000 \text{ mm}}{1 \text{ m}}\right) = 1000 \text{ m/s}^2$$

(b) Having learned the acceleration from part (a) we can now rearrange the first equation in Table 2.4 to find the time it takes to reach takeoff speed. Again use $(v_y)_i = 0.0$ m/s.

$$\Delta t = \frac{(v_y)_f}{a_y} = \frac{1.0 \text{ m/s}}{1000 \text{ m/s}^2} = .0010 \text{ s}$$

(c) This time $(v_y)_f = 0.0$ m/s as the flea reaches the top of its trajectory. Rearrange the last equation in Table 2.4 to get

$$\Delta y = \frac{-(v_y)_i^2}{2a_y} = \frac{-(1.0 \text{ m/s})^2}{2(-9.8 \text{ m/s}^2)} = 0.051 \text{ m} = 5.1 \text{ cm}$$

Assess: Just over 5 cm is pretty good considering the size of a flea. It is about 10–20 times the size of a typical flea.
Check carefully to see that each answer ends up in the appropriate units.
The height of the flea at the top will round to 5.2 cm above the ground if you include the 0.050 cm during the initial acceleration phase before the feet actually leave the ground.

P2.63. Prepare: A visual overview of the rock's motion that includes a pictorial representation, a motion diagram, and a list of values is shown below. We represent the rock's motion along the y-axis. As soon as the rock is tossed up, it falls freely and thus kinematic equations hold. The rock's acceleration is equal to the acceleration due to gravity that always acts vertically downward toward the center of the earth. The initial position of the rock is at the origin where $y_i = 0$, but the final position is below the origin at $y_f = -10$ m. Recall sign conventions which tell us that v_i is positive and a is negative.

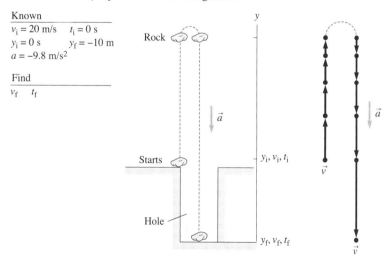

Solve: (a) Substituting the known values into $y_f = y_i + v_i \Delta t + \frac{1}{2} a \Delta t^2$, we get

$$-10 \text{ m} = 0 \text{ m} + 20 \text{ (m/s)}t_f + \frac{1}{2}(-9.8 \text{ m/s}^2)t_f^2$$

One of the roots of this equation is negative and is not physically relevant. The other root is $t_f = 4.53$ s which is the answer to part (b). Using $v_f = v_i + a \Delta t$, we obtain

$$v_f = 20 (\text{m/s}) + (-9.8 \text{ m/s}^2)(4.53 \text{ s}) = -24 \text{ m/s}$$

(b) The time is 4.5 s.
Assess: A time of 4.5 s is a reasonable value. The rock's velocity as it hits the bottom of the hole has a negative sign because of its downward direction. The magnitude of 24 m/s compared to 20 m/s when the rock was tossed up is consistent with the fact that the rock travels an additional distance of 10 m into the hole.

P2.65. Prepare: It is clear the second ball must be thrown with a greater initial velocity than the first ball in order to catch up to it at the top. In the Assess step we will verify that this is indeed the case. We are *not* told the second ball has zero velocity when it hits the first ball at the top of the first ball's trajectory (the first ball *would* have zero velocity at this time).
There are several steps in this problem, so first draw a picture and, like the examples in the book, list the known quantities and what we need to find.
Call the juggler's hand the origin and call $t = 0$ s when the first ball is released. We will assume both stones are in free fall after they leave the juggler's hand, so $a_y = -g$. Let a subscript 1 refer to the first ball and a 2 refer to the second.

Known
$(y_1)_i = 0.0 \text{ m}$
$(y_2)_i = 0.0 \text{ m}$
$(y_2)_f = (y_1)_f$; simply call this y_r
$(v_1)_i = 10 \text{ m/s}$
$(v_1)_f = 0.0 \text{ m/s}$
$(t_1)_i = 0.0 \text{ s}$
$(t_2)_i = 0.50 \text{ s}$
$(t_2)_f = (t_1)_f$; simply call this t_r

Find
$(v_2)_i$

The strategy will be to find the position and time of the first ball at the top of its trajectory, and then to compute the initial speed needed for the second ball to get to the same place at the same time.
Solve: Use the first equation of Table 2.4 to find t_f (which is equal to $(t_1)_i + \Delta t_1$). Everything is in the y direction so we drop the y subscript.

$$(v_1)_f = (v_1)_i + a \Delta t_1$$

Use the facts that $(v_1)_f = 0.0$ m/s and $a = -g$ to solve for Δt_1.

$$\Delta t_1 = \frac{-(v_1)_i}{-g} = \frac{10 \text{ m/s}}{9.8 \text{ m/s}^2} = 1.02 \text{ s}$$

Then use $(t_1)_i = 0.0$ s to find $t_f = 1.02$ s.

Now use the third equation of Table 2.4 to find the position of the top of the first ball's trajectory.

$$(v_1)_f^2 = (v_1)_i^2 + 2a\Delta y$$

We know that $(v_1)_f = 0.0$ m/s and $a = -g$.

$$\Delta y = \frac{-(v_1)_i^2}{2a} = \frac{-(10.0 \text{ m/s})^2}{2(-g)} = 5.1 \text{ m}$$

Since $(y_1)_i = 0.0$ m we know that $(y_1)_f = y_f = 5.1$ m

Those were the preliminaries. Now we use $t_f = 1.02$ s and $y_f = 5.1$ m in the second equation of Table 2.4 to solve for $(v_2)_i$. For the second ball $(t_2)_i = 0.5$ s so $\Delta t_2 = 1.02 \text{ s} - 0.50 \text{ s} = 0.52$ s.

$$y_f = (y_2)_i + (v_2)_i \Delta t_2 + \frac{1}{2} a (\Delta t_2)^2$$

Solving for $(v_2)_i$:

$$(v_2)_i = \frac{y_f - (y_2)_i - \frac{1}{2} a (\Delta t_2)^2}{\Delta t_2} = \frac{5.1 \text{ m} - 0.0 \text{ m} - \frac{1}{2}(-g)(0.52 \text{ s})^2}{0.52 \text{ s}} = \frac{5.1 \text{ m} + \frac{1}{2}(9.8 \text{ m/s}^2)(0.52 \text{ s})^2}{0.52 \text{ s}} = 12 \text{ m/s}$$

Assess: Our original statement that $(v_2)_i$ had to be greater than $(v_1)_i$ is correct. The answer still seems to be a reasonable throwing speed.

P2.71. Prepare: Before we turn to algebra, carefully examine the velocity-versus-time graph below. We draw the line for car 1 by starting at the origin at time zero and making the slope 2.0 m/s². The line for car 2 starts at $v_2 = 0.0$ m/s when $t = 2.0$ s and has a slope of 8.0 m/s². The time where the vertical dotted line should be placed is initially unknown; think of sliding it left and right until the areas of the two triangles (under the graphs of cars 1 and 2) are the same. When that happens then the time where the dotted line ends up is the answer to part (**a**), and the area under the two lines (*i.e.*, the area of each triangle) will be the answer to part (**b**).

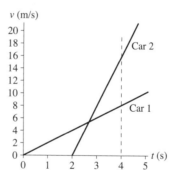

Since the accelerations are constant the equations in Table 2.4 will apply. Assume that each car starts from rest. Call the launch point the origin and call $t = 0$s when the first car is launched. Let a subscript 1 refer to the first car and a 2 refer to the second.

Known
_____

$(x_1)_i = 0.0 \text{ m}$

$(x_2)_i = 0.0 \text{ m}$

$(x_2)_f = (x_1)_f;$ simply call this x_f

$(v_1)_i = 0.0 \text{ m/s}$

$(v_2)_i = 0.0 \text{ m/s}$

$(t_1)_i = 0.0 \text{ s}$

$(t_2)_i = 2.0 \text{ s}$

$(t_2)_f = (t_1)_f;$ simply call this t_f

$a_1 = 2.0 \text{ m/s}^2$

$a_2 = 8.0 \text{ m/s}^2$

Find
_____

t_f

x_f

The strategy will be to use equations from Table 2.4 with the twist that no one of them has only one unknown (since both final velocities are unknown). So we will use various instances of the equations to get a system of equations with enough equations to solve for our unknowns.

Solve: Apply the last equation in Table 2.4 to each car (with the initial velocities both zero) and then divide the equations. Dividing equations to get ratios is an elegant and useful technique.

$$(v_1)_f^2 = 2a_1 \Delta x$$

$$(v_2)_f^2 = 2a_2 \Delta x$$

$$\left(\frac{(v_1)_f}{(v_2)_f} \right)^2 = \frac{a_1}{a_2} = \frac{2.0 \text{ m/s}^2}{8.0 \text{ m/s}^2} = \frac{1}{4}$$

Taking square roots gives

$$\frac{(v_1)_f}{(v_2)_f} = \frac{1}{2}$$

(a) Now turn to the first equation in Table 2.4 and apply it to both cars (again, with both initial velocities zero). Since $(t_1)_i = 0.0 \text{s}$ then $\Delta t_1 = t_f$, and since $(t_2)_i = 2.0 \text{ s}$ then $\Delta t_2 = t_f - 2.0 \text{ s}$,

$$(v_1)_f = a_1 t_f \qquad (v_2)_f = a_2 (t_f - 2.0 \text{ s})$$

Divide the equations and use the previous result.

$$\frac{(v_1)_f}{(v_2)_f} = \frac{a_1}{a_2} \left(\frac{t_f}{t_f - 2.0 \text{ s}} \right) \frac{1}{2} = \frac{1}{4} \left(\frac{t_f}{t_f - 2.0 \text{ s}} \right)$$

Solve for t_f by multiplying both sides by $4(t_f - 2.0)$.

$$2(t_f - 2.0 \text{ s}) = t_f \qquad 2t_f - 4.0 \text{ s} = t_f \qquad 2t_f - t_f = 4.0 \text{ s} \qquad t_f = 4.0 \text{ s}$$

This is the answer to part **(a)**: $t_f = 4.0 \text{ s}$.

(b) Although we aren't asked for the final velocities, we compute $(v_1)_f$ so can get find the distance traveled.

$$(v_1)_f = a_1 t_f = (2.0 \text{ m/s}^2)(4.0 \text{ s}) = 8.0 \text{ m/s}$$

With $(x_1)_i = 0.0 \text{ m}$ then $\Delta x_1 = x_f$. Since $(v_1)_f = 0.0 \text{ m/s}$, the last equation in Table 2.4 becomes for car 1

$$(v_1)_f^2 = 2a_1 x_f$$

So

$$x_f = \frac{(v_1)_f^2}{2a_1} = \frac{(8.0 \text{ m/s})^2}{2(2.0 \text{ m/s}^2)} = 16 \text{ m}$$

A similar calculation for car 2 gives the same result; the two cars meet 16 m down the track.
Assess: We have finally answered both parts of the problem. We check that the units are correct.
We note the important technique of dividing one equation by another to get dimensionless ratios.
Last, we note that the algebra agrees with the graphical approach taken at the beginning.

P2.73. Prepare: A visual overview of the car's motion that includes a pictorial representation, a motion diagram, and a list of values is shown below. We label the car's motion along the x-axis. This is a two-part problem. First, we need to use the information given to determine the acceleration during braking. We will then use this acceleration to find the stopping distance for a different initial velocity.

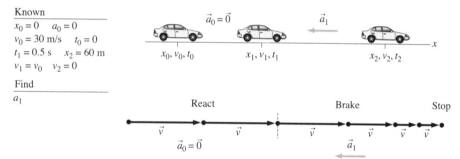

Solve: **(a)** First, the car coasts at constant speed before braking:

$$x_1 = x_0 + v_0(t_1 - t_0) = v_0 t_1 = (30 \text{ m/s})(0.5 \text{ s}) = 15 \text{ m}$$

Then, the car brakes to a halt. Because we don't know the time interval during braking, we will use

$$v_2^2 = 0 = v_1^2 + 2a_1(x_2 - x_1)$$

$$\Rightarrow a_1 = -\frac{v_1^2}{2(x_2 - x_1)} = -\frac{(30 \text{ m/s})^2}{2(60 \text{ m} - 15 \text{ m})} = -10 \text{ m/s}^2$$

We use $v_1 = v_0 = 30 \text{ m/s}$. Note the minus sign, because $\vec{a}_1$ points to the left. We can repeat these steps now with $v_0 = 40 \text{ m/s}$. The coasting distance before braking is

$$x_1 = v_0 t_1 = (40 \text{ m/s})(0.5 \text{ s}) = 20 \text{ m}$$

So the braking distance is

$$v_2^2 = 0 = v_1^2 + 2a_1(x_2 - x_1)$$

$$\Rightarrow x_2 = x_1 - \frac{v_1^2}{2a_1} = 20 \text{ m} - \frac{(40 \text{ m/s})^2}{2(-10 \text{ m/s}^2)} = 100 \text{ m}$$

(b) The car coasts at a constant speed for 0.5 s, traveling 20 m. The graph will be a straight line with a slope of 40 m/s. For $t \geq 0.5$ the graph will be a parabola until the car stops at t_2. We can find t_2 from

$$v_2 = 0 = v_1 + a_1(t_2 - t_1) \Rightarrow t_2 = t_1 - \frac{v_1}{a_1} = 4.5 \text{ s}$$

The parabola will reach zero slope ($v = 0$ m/s) at $t = 4.5$ s. This is enough information to draw the graph shown in the figure.

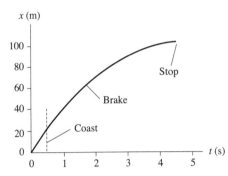

3

Q3.3. Reason: Consider two vectors $\vec{A}$ and $\vec{B}$. Their sum can be found using the method of algebraic addition. In Question 3.2 we found that the components of the zero vector are both zero. The components of the resultant of $\vec{A}$ and $\vec{B}$ must then be zero also. So

$$R_x = A_x + B_x = 0$$
$$R_y = A_y + B_y = 0$$

Solving for the components of $\vec{B}$ in terms of $\vec{A}$ gives $B_x = -A_x$ and $B_y = -A_y$. Then the magnitude of $\vec{B}$ is $\sqrt{(B_x)^2 + (B_y)^2} = \sqrt{(-A_x)^2 + (-A_y)^2} = \sqrt{(A_x)^2 + (A_y)^2}$. So then the magnitude of $\vec{B}$ is exactly equal to the magnitude of $\vec{A}$.

Assess: For two vectors to add to zero, the vectors must have exactly the same magnitude and point in opposite directions.

Q3.5. Reason: The ones that are constant are v_x, a_x, and a_y. Furthermore, a_x is not only constant, it is zero.

Assess: There are instants when other quantities can be zero, but not throughout the flight. Remember that $a_y = -g$ throughout the flight and that v_x is constant; that is, projectile motion is nothing more than the combination of two simple kinds of motion: constant horizontal velocity and constant vertical acceleration.

Q3.9. Reason: The claim is slightly misleading, since the passenger cannot walk at a speed in excess of 500 mph due to her own efforts relative to the walking surface (the floor of the plane in this case); but she can be walking and moving with such a speed relative to the ground thousands of feet below.

Assess: It is important to specify the coordinate system when reporting velocities.

Q3.11. Reason: The acceleration is due to gravity, so the acceleration will always act to pull the cars back down the ramp. Since the roller coaster is constrained to move along the ramp, the acceleration must be along the ramp. So in all three cases the acceleration is downward along the ramp.

Assess: Gravity always acts, even if motion is constrained by an inclined ramp. See Figure 3.24 in the text.

Q3.15. Reason: Since the plane is moving in a circle and constantly changing direction, the plane is constantly accelerating. At any point along its motion, even directly north, the plane is accelerating. For circular motion, the direction of the acceleration is always toward the center of the circle. When the plane is headed directly north, the plane's acceleration vector points directly east. See the next figure.

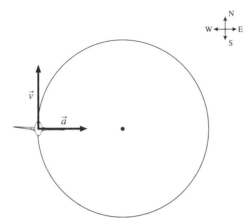

Assess: In circular motion, there is a centripetal acceleration that always points toward the center of the circle.

Problems

P3.1. Prepare: (a) To find $\vec{A} + \vec{B}$, we place the tail of vector $\vec{B}$ on the tip of vector $\vec{A}$ and then connect vector $\vec{A}$'s tail with vector $\vec{B}$'s tip. (b) To find $\vec{A} - \vec{B}$, we note that $\vec{A} - \vec{B} = \vec{A} + (-\vec{B})$. We place the tail of vector $-\vec{B}$ on the tip of vector $\vec{A}$ and then connect vector $\vec{A}$'s tail with the tip of vector $-\vec{B}$.
Solve:

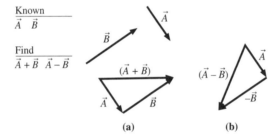

P3.3. Prepare: We can find the positions and velocity and acceleration vectors using a motion diagram.

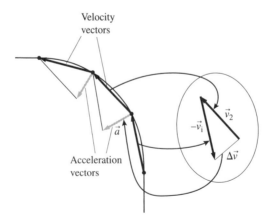

Solve: The figure gives several points along the car's path. The velocity vectors are obtained by connecting successive dots. The acceleration vectors are obtained by subtracting successive velocity vectors. The acceleration vectors point toward the center of the diagram.
Assess: Notice that the acceleration points toward the center of the turn. As you will learn in chapter 4, whenever your car accelerates, you feel like you are being pushed the opposite way. This is why you feel like you are being pushed away from the center of a turn.

P3.7.   Prepare: The figure below shows the components v_x and v_y, and the angle θ. We will use Tactics Box 3.3 to find the sign attached to the components of a vector.

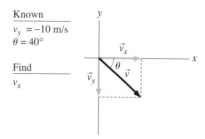

Solve: We have,

$$v_y = -v \sin 40°, \quad \text{or} \quad -10 \text{ m/s} = -v \sin 40°, \quad \text{or} \quad v = 15.56 \text{ m/s.}$$

Thus the *x*-component is $v_x = v \cos 40° = (15.56 \text{ m/s}) \cos 40° = 12 \text{ m/s.}$

Assess: Note that we had to insert the minus sign manually with v_y since the vector is in the fourth quadrant.

P3.9.   Prepare: The figure below shows the components $v_{||}$ and $v_{\perp}$, and the angle θ. We will use Tactics Box 3.3 to find the sign attached to the components of a vector.

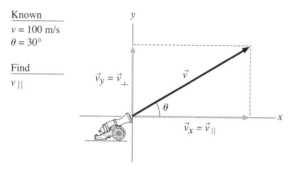

Solve: We have $\vec{v} = \vec{v}_x + \vec{v}_y = \vec{v}_{||} + \vec{v}_{\perp}$. Thus, $v_{||} = v \cos \theta = (100 \text{ m/s}) \cos 30° = 87 \text{ m/s.}$

Assess: For the small angle of 30°, the obtained value of 87 m/s for the horizontal component is reasonable.

P3.11.   Prepare: We will follow rules given in the Tactics Box 3.3.

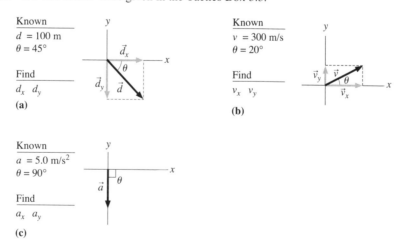

Solve: **(a)** Vector $\vec{d}$ points to the right and down, so the components d_x and d_y are positive and negative, respectively:

$$d_x = d \cos \theta = (100 \text{ m}) \cos 45° = 70.7 \text{ m} \qquad d_y = -d \sin \theta = -(100 \text{ m}) \sin 45° = -71 \text{ m}$$

(b) Vector $\vec{v}$ points to the right and up, so the components v_x and v_y are both positive:

$$v_x = v \cos \theta = (300 \text{ m/s}) \cos 20° = 280 \text{ m/s} \qquad v_y = v \sin \theta = (300 \text{ m/s}) \sin 20° = 100 \text{ m/s}$$

(c) Vector $\vec{a}$ has the following components:

$$a_x = -a \cos \theta = -(5.0 \text{ m/s}^2) \cos 90° = 0 \text{ m/s}^2 \qquad a_y = -a \sin \theta = -(5.0 \text{ m/s}^2) \sin 90° = -5.0 \text{ m/s}^2$$

Assess: The components have same units as the vectors. Note the minus signs we have manually inserted according to the Tactics Box 3.3.

P3.13. Prepare: We will draw the vectors to scale as best we can and label the angles from the positive *x*-axis (positive angles go CCW). We also use Equations 3.11 and 3.12. Make sure your calculator is in degree mode.
Solve: **(a)**

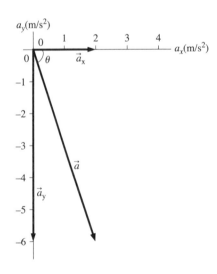

$$v = \sqrt{(v_x)^2 + (v_y)^2} = \sqrt{(20 \text{ m/s})^2 + (40 \text{ m/s})^2} = 45 \text{ m/s}$$

$$\theta = \tan^{-1}\left(\frac{v_y}{v_x}\right) = \tan^{-1}\left(\frac{40 \text{ m/s}}{20 \text{ m/s}}\right) = \tan^{-1}(2) = 63°$$

(b)

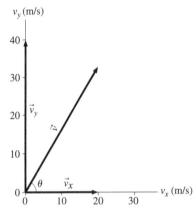

$$a = \sqrt{(a_x)^2 + (a_y)^2} = \sqrt{(2.0 \text{ m/s}^2)^2 + (-6.0 \text{ m/s}^2)^2} = 6.3 \text{ m/s}^2$$

$$\theta = \tan^{-1}\left(\frac{a_y}{a_x}\right) = \tan^{-1}\left(\frac{-6.0 \text{ m/s}^2}{2.0 \text{ m/s}^2}\right) = \tan^{-1}(-3) = -72°$$

Assess: In each case the magnitude is longer than either component, as is required for the hypotenuse of a right triangle.

The negative angle in part (**b**) corresponds to a clockwise direction from the positive *x*-axis.

P3.19. Prepare: Make a sketch with tilted axes with the *x*-axis parallel to the ramp and the angle of inclination labeled. We must also make a bold assumption that the piano rolls down as if it were an object sliding down with no friction.

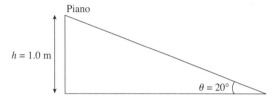

As part of the preparation, compute the length of the ramp in the new tilted *x-y* coordinates.

$$L = \frac{h}{\sin \theta} = \frac{1.0 \text{ m}}{\sin 20°} = 2.9 \text{ m}$$

The acceleration in the new coordinate system will be $a_x = g \sin \theta = (9.8 \text{ m/s}^2) \sin 20° = 3.4 \text{ m/s}^2$.

Solve: Since this is a case of constant acceleration we can use the second equation from Table 2.4 with $x_i = 0.0$ m and $(v_x)_i = 0.0$ m/s.

$$x_f = \frac{1}{2} a_x (\Delta t)^2$$

Solve for Δt, and use $x_f = 2.9$ m and $a_x = 3.4 \text{ m/s}^2$, which we obtained previously.

$$\Delta t = \sqrt{\frac{2x_f}{a_x}} = \sqrt{\frac{2(2.9 \text{ m})}{3.4 \text{ m/s}^2}} = 1.3 \text{ s}$$

Assess: They may catch it if they have quick reactions, but the piano will be moving 4.5 m/s when it reaches the bottom.

P3.21. Prepare: For everyday speeds we can use Equation 3.22 to find relative velocities. We will use a subscript A for Anita and a 1 and a 2 for the respective balls; we also use a subscript G for the ground. We will consider all motion in this problem to be along the *x*-axis (ignore the vertical motion including the fact that the balls also fall under the influence of gravity) and so we drop the *x* subscript.

It is also worth noting that interchanging the order of the subscripts merely introduces a negative sign. For example, $v_{AG} = 5$ m/s, so $v_{GA} = -5$ m/s.

"According to Anita" means "relative to Anita."

Solve: For ball 1:

$$v_{1A} = v_{1G} + v_{GA} = 10 \text{ m/s} + (-5 \text{ m/s}) = 5 \text{ m/s}$$

For ball 2:

$$v_{2A} = v_{2G} + v_{GA} = -10 \text{ m/s} + (-5 \text{ m/s}) = -15 \text{ m/s}$$

The speed is the magnitude of the velocity, so the speed of ball 2 is 15 m/s.

Assess: You can see that at low speeds velocities simply add or subtract, as the case may be. Mentally put yourself in Anita's place, and you will confirm that she sees ball 1 catching up to her at only 5 m/s while she sees ball 2 speed past her at 15 m/s.

P3.27. Prepare: We will assume the ball is in free fall (*i.e.*, we neglect air resistance). The trajectory of a projectile is a parabola because it is a combination of constant horizontal velocity ($a_x = 0.0$ m/s^2) combined with constant vertical acceleration ($a_y = -g$). In this case we see only half of the parabola.

The initial speed given is all in the horizontal direction, that is, $(v_x)_i = 5.0$ m/s and $(v_y)_i = 0.0$ m/s.

Solve:

(a)   **(b)**   **(c)**

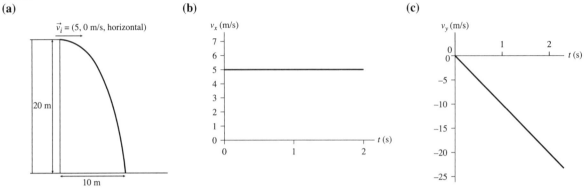

(d) This is a two-step problem. We first use the vertical direction to determine the time it takes, then plug that result into the equation for the horizontal direction.

$$\Delta y = \frac{1}{2}a_y(\Delta t)^2$$

$$\Delta t = \sqrt{\frac{2\Delta y}{a_y}} = \sqrt{\frac{2(-20 \text{ m})}{-9.8 \text{ m/s}^2}} = 2.0 \text{ s}$$

We we use the 2.0 s in the equation for the horizontal motion.

$$\Delta x = v_x \Delta t = (5.0 \text{ m/s})(2.0 \text{ s}) = 10 \text{ m}$$

Assess: The answers seem reasonable, and we would get the same answers to two significant figures in a quick mental calculation using $g \approx 10$ m/s^2. In fact, I did this before computing the algebra so I would know how to scale the graphs.

P3.31. Prepare: We will apply the constant-acceleration kinematic equations to the horizontal and vertical motions as described by Equations 3.25. The effect of air resistance on the motion of the bullet is neglected.

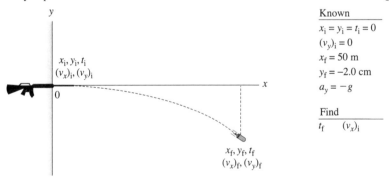

Known
$x_i = y_i = t_i = 0$
$(v_y)_i = 0$
$x_f = 50$ m
$y_f = -2.0$ cm
$a_y = -g$

Find
t_f $(v_x)_i$

Solve: (a) Using $y_f = y_i + (v_y)_i(t_f - t_i) + \frac{1}{2}a_y(t_f - t_i)^2$, we obtain

$$(-2.0 \times 10^{-2} \text{ m}) = 0 \text{ m} + 0 \text{ m} + \tfrac{1}{2}(-9.8 \text{ m/s}^2)(t_f - 0 \text{ s})^2 \Rightarrow t_f = 0.0639 \text{ s}$$

(b) Using $x_f = x_i + (v_x)_i(t_f - t_i) + \frac{1}{2}a_x(t_f - t_i)^2$,

$$(50 \text{ m}) = 0 \text{ m} + (v_x)_i(0.0639 \text{ s} - 0 \text{ s}) + 0 \text{ m} \Rightarrow (v_x)_i = 782 \text{ m/s}$$

Assess: The bullet falls 2 cm during a horizontal displacement of 50 m. This implies a large initial velocity, and a value of 782 m/s is not surprising.

P3.35. Prepare: We need to convert the 5400 rpm to different units and then find the period which is the inverse of frequency.

Solve: (a) The hard disk's frequency can be converted as follows:

$$5400 \frac{\text{rev}}{\text{min}} = 5400 \frac{\text{rev}}{\text{min}} \left(\frac{1 \text{ min}}{60 \text{ sec}} \right) = 90 \frac{\text{rev}}{\text{sec}}$$

Its frequency is 90 rev/s.

(b) Rewriting Equation 3.26, we have the following:

$$T = \frac{1}{f} = \frac{1}{90 \text{ rev/s}} = 11 \text{ ms}$$

Its period is 11 ms.

Assess: This is about the rate that the engine in a car turns if it is straining. So an automobile engine completes a cycle every 10 or 20 ms.

P3.37. Prepare: We are asked to find period, speed and acceleration. Period and frequency are inverses according to Equation 3.26. To find speed we need to know the distance traveled by the speck in one period. Then the acceleration is given by Equation 3.30.

Solve: **(a)** The disk's frequency can be converted as follows:

$$10,000 \frac{\text{rev}}{\text{min}} = 10,000 \frac{\text{rev}}{\text{min}} \left(\frac{1 \text{ min}}{60 \text{ sec}} \right) = 167 \frac{\text{rev}}{\text{sec}}$$

The period is the inverse of the frequency:

$$T = \frac{1}{f} = \frac{1}{167 \text{ rev/s}} = 6.00 \text{ ms}$$

(b) The speed of the speck equals the circumference of its orbit divided by the period:

$$v = \frac{2\pi r}{T} = \frac{2\pi (6.0 \text{ cm})}{6.00 \text{ ms}} \left(\frac{1000 \text{ ms}}{1 \text{ s}} \right) \left(\frac{1 \text{ m}}{100 \text{ cm}} \right) = 62.8 \text{ m/s},$$

which rounds to 63 m/s.

(c) From Equation 3.30, the acceleration of the speck is given by v^2 / r:

$$a = \frac{v^2}{r} = \frac{(62.8 \text{ m/s})^2}{6.0 \text{ cm}} \left(\frac{100 \text{ cm}}{1 \text{ m}} \right) = 65,700 \text{ m/s}^2,$$

Which rounds to 66,000 m/s². In units of *g*, this is as follows:

$$65,700 \text{ m/s}^2 = 65,700 \text{ m/s}^2 \left(\frac{1g}{9.8 \text{ m/s}^2} \right) = 6,700g$$

Assess: The speed and acceleration of the edge of a CD are remarkable. The speed, 63 m/s, is about 140 mi/hr. As you will learn in chapter 4, very large forces are necessary to create large accelerations like 6,700*g*.

P3.43. Prepare: The vectors $\vec{A}$, $\vec{B}$, and $\vec{D} = \vec{A} - \vec{B}$ are shown. Because $\vec{A} = \vec{A}_x + \vec{A}_y$ and $\vec{B} = \vec{B}_x + \vec{B}_y$, so the components of the resultant vector are $D_x = A_x - B_x$ and $D_y = A_y - B_y$.

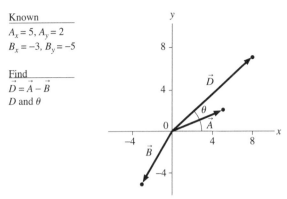

Solve: (a) With $A_x = 5$, $A_y = 2$, $B_x = -3$, and $B_y = -5$, we have $D_x = 8$ and $D_y = 7$.

(b) Vectors $\vec{A}$, $\vec{B}$ and $\vec{D}$ are shown in the above figure.

(c) Since $D_x = 8$ and $D_y = 7$, the magnitude and direction of $\vec{D}$ are

$$D = \sqrt{(8)^2 + (7)^2} = 11 \qquad \theta = \tan^{-1}\left(\frac{D_y}{D_x}\right) = \tan^{-1}\left(\frac{7}{8}\right) = 41°$$

Assess: Since $|D_y| < |D_x|$, the angle θ is less than 45°, as it should be.

P3.45. Prepare: Refer to Figure P3.45 in your textbook. Because $\vec{A} = \vec{A}_x + \vec{A}_y$, $\vec{B} = \vec{B}_x + \vec{B}_y$, and $\vec{C} = \vec{C}_x + \vec{C}_y$, so the components of the resultant vector are $D_x = A_x + B_x + C_x$ and $D_y = A_y + B_y + C_y$. D_x and D_y are given and we will read the components of $\vec{A}$ and $\vec{C}$ off Figure P3.45.

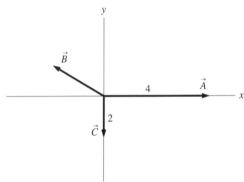

Solve: (a) $A_x = 4$, $C_x = 0$, and $D_x = 2$, so $B_x = A_x - C_x + D_x = -2$. Similarly, $A_y = 0$, $C_y = -2$, and $D_y = 0$, so $B_y = -A_y - C_y + D_y = 2$.

(b) With the components in (a), $B = \sqrt{(-2)^2 + (2)^2} = 2.8$

$$\theta = \tan^{-1}\frac{B_y}{|B_x|} = \tan^{-1}\frac{2}{2} = 45°$$

Since $\vec{B}$ has a negative x-component and a positive y-component, the angle θ made by $\vec{B}$ is with the $-x$-axis and it is above the $-x$-axis.

Assess: Since $|B_y| = |B_x|$, $\theta = 45°$ as is obtained above.

P3.49. Prepare: In the coordinate system shown below, the mouse starts from the origin, so his initial position vector is zero. His net displacement is then just the final position vector, which is just the sum of the three vectors $\vec{A}$, $\vec{B}$, and $\vec{C}$.

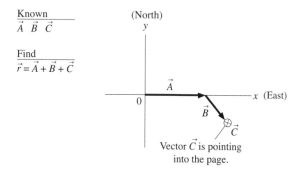

Solve: We are given $\vec{A} = (5 \text{ m, east})$ and $\vec{C} = (1 \text{ m, down})$. Using trigonometry,

$\vec{B} = (3 \cos 45° \text{ m, east}) + (3 \sin 45° \text{ m, south})$.

The total displacement is $\vec{r} = \vec{A} + \vec{B} + \vec{C} = (7.12 \text{ m, east}) + (2.12 \text{ m, south}) + (1 \text{ m, down})$.
The magnitude of $\vec{r}$ is

$$r = \sqrt{(7.12)^2 + (2.12)^2 + (1)^2} \text{ m} = 7.5 \text{ m}$$

Assess: A displacement of 7.5 m is a reasonable displacement.

P3.53. Prepare: The skier's motion on the horizontal, frictionless snow is not of any interest to us. The skier's speed increases down the incline due to acceleration parallel to the incline, which is equal to $g \sin 10°$. A visual overview of the skier's motion that includes a pictorial representation, a motion representation, and a list of values is shown.

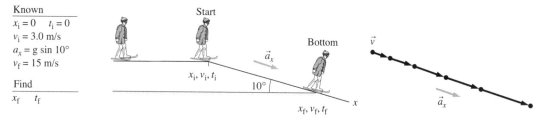

Solve: Using the following constant-acceleration kinematic equations,

$$v_f^2 = v_i^2 + 2a_x(x_f - x_i)$$
$$\Rightarrow (15 \text{ m/s})^2 = (3.0 \text{ m/s})^2 + 2(9.8 \text{ m/s}^2) \sin 10°(x_f - 0 \text{ m}) \Rightarrow x_f = 64 \text{ m}$$
$$v_f = v_i + a_x(t_f - t_i)$$
$$\Rightarrow (15 \text{ m/s}) = (3.0 \text{ m/s}) + (9.8 \text{ m/s}^2)(\sin 10°)t_f \Rightarrow t_f = 7.1 \text{ s}$$

Assess: A time of 7.1 s to cover 64 m is a reasonable value.

P3.55. Prepare: A visual overview of the puck's motion that includes a pictorial representation, a motion diagram, and a list of values is shown as follows.

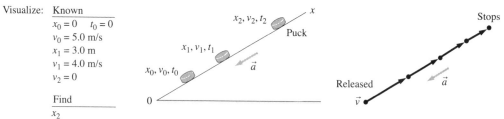

Solve: The acceleration, being the same along the incline, can be found as

$$v_1^2 = v_0^2 + 2a(x_1 - x_0) \Rightarrow (4.0 \text{ m/s})^2 = (5.0 \text{ m/s})^2 + 2a(3.0 \text{ m} - 0 \text{ m}) \Rightarrow a = -1.5 \text{ m/s}^2$$

We can also find the total time the puck takes to come to a halt as

$$v_2 = v_0 + a(t_2 - t_0) \Rightarrow 0 \text{ m/s} = (5.0 \text{ m/s}) + (-1.5 \text{ m/s}^2)t_2 \Rightarrow t_2 = 3.3 \text{ s}$$

Using the above obtained values of a and t_2, we can find x_2 as follows:

$$x_2 = x_0 + v_0(t_2 - t_0) + \frac{1}{2}a(t_2 - t_0)^2 = 0 \text{ m} + (5.0 \text{ m/s})(3.3 \text{ s}) + \frac{1}{2}(-1.5 \text{ m/s}^2)(3.3 \text{ s})^2 = 8.3 \text{ m}$$

That is, the puck goes through a displacement of 8.3 m. Since the end of the ramp is 8.5 m from the starting position x_0 and the puck stops 0.2 m or 20 cm before the ramp ends, you are not a winner.

P3.57. Prepare: Both ships have a common origin at $t = 0$ s.
Solve: (a) The velocity vectors of the two ships are as follows:

$$\vec{v}_A = (20 \text{ mph} \cos 30°, \text{ north}) + (20 \text{ mph} \sin 30°, \text{ west}) = (17.32 \text{ mph}, \text{ north}) + (10.0 \text{ mph}, \text{ west})$$
$$\vec{v}_B = (25 \text{ mph} \cos 20°, \text{ north}) + (25 \text{ mph} \sin 20°, \text{ east}) = (23.49 \text{ mph}, \text{ north}) + (8.55 \text{ mph}, \text{ east})$$

Since $\vec{r} = \vec{v}\Delta t$,

$$\vec{r}_A = \vec{v}_A(2 \text{ hr}) = (34.64 \text{ mph}, \text{ north}) + (20.0 \text{ mph}, \text{ west})$$
$$\vec{r}_B = \vec{v}_B(2 \text{ hr}) = (46.98 \text{ miles}, \text{ north}) + (17.10 \text{ miles}, \text{ east})$$

So, the displacement between the two ships is

$$\vec{R} = \vec{r}_A - \vec{r}_B = (12.34 \text{ miles}, \text{ south}) + (37.10 \text{ miles}, \text{ west}) \Rightarrow R = 39.1 \text{ miles}$$

(b) The speed of A as seen by B is: $\vec{V} = \vec{v}_A - \vec{v}_B = (6.17 \text{ mph}, \text{ south}) + (18.55 \text{ mph}, \text{ west}) \Rightarrow V = 19.5 \text{ mph}$.
Assess: The value of the speed is reasonable.

P3.61. Prepare: A visual overview of the plane's motion is shown in the following figure. The direction the pilot must head the plane can be obtained from $\vec{v}_{pg} = \vec{v}_{pa} + \vec{v}_{ag}$, where $\vec{v}_{pa} = (v_{pa} \sin \theta, \text{ south}) + (v_{pa} \cos \theta, \text{ east})$, $v_{pa} = 200 \text{ mph}$, $\vec{v}_{ag} = (v_{ag} \sin 30°, \text{ north}) + (v_{ag} \cos 30°, \text{ east})$, and $\vec{v}_{pg} = (v_{pg}, \text{ east})$.

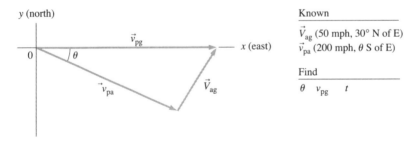

Solve: (a) Writing the equation $\vec{v}_{pg} = \vec{v}_{pa} + \vec{v}_{ag}$ in the form of components

$$(v_{pg}, \text{ east}) = [\,(200 \text{ mph} \sin \theta, \text{ south}) + (200 \text{ mph} \cos \theta, \text{ east})\,] +$$
$$[\,(50 \text{ mph} \sin 30°, \text{ north}) + (50 \text{ mph} \cos 30°, \text{ east})\,]$$

Because $(\vec{v})_{pg}$ should have no component along north,

$$50 \sin 30° - 200 \sin \theta = 0 \Rightarrow \theta = 7.2°$$

(b) The pilot must head 7.18° south of east. Substituting this value of θ in the above velocity equation gives $(v_{pg}, \text{east}) = (200 \text{ mph} \cos 7.18°, \text{east}) + (50 \text{ mph} \cos 30°, \text{east}) = (240 \text{ mph}, \text{east})$. At a speed of 240 mph, the trip takes $t = 600 \text{ mi}/240 \text{ mph} = 2.5$ hours.

P3.65. Prepare: This problem is somewhat similar to Problem 3.27 with all of the initial velocity in the horizontal direction. We will use the vertical equation for constant acceleration to determine the time of flight and then see how far Captain Brady can go in that time.
Of interest is the fact that we will do this two-step problem completely with variables in part **(a)** and only plug in numbers in part **(b)**.
We *could* do part **(b)** in feet (using $g = 32 \text{ ft/s}^2$), but to compare with the world record 100 m dash, let's convert to meters. $L = 22 \text{ ft} = 6.71 \text{ m}$ and $h = 20 \text{ ft} = 6.10 \text{ m}$.
Solve: (a) Given that $(v_y) = 0.0 \text{ ft/s}$ we can use Equation 2.14.

$$\Delta y = \frac{1}{2} a_y (\Delta t)^2$$

With up as the positive direction, Δy is negative and $a_y = -g$; those signs cancel leaving

$$h = \frac{1}{2} g (\Delta t)^2$$

Solve for Δt.

$$\Delta t = \sqrt{\frac{2h}{g}}$$

Now use that expression for Δt in the equation for constant horizontal velocity.

$$L = \Delta x = v_x \Delta t = v_x \sqrt{\frac{2h}{g}}$$

Finally solve for $v = v_x$ in terms of L and h.

$$v = \frac{L}{\sqrt{\frac{2h}{g}}} = L \sqrt{\frac{g}{2h}}$$

(b) Now plug in the numbers we are given for L and h.

$$v = L \sqrt{\frac{g}{2h}} = (6.71 \text{ m}) \sqrt{\frac{9.8 \text{ m/s}^2}{2(6.10 \text{ m})}} = 6.0 \text{ m/s}$$

Compare this result ($v = 6.0 \text{ m/s}$) with the world-class sprinter ($v = 10 \text{ m/s}$); a fit person could make this leap.
Assess: The results are reasonable, and not obviously wrong. $6.0 \text{ m/s} \approx 13 \text{ mph}$, and that would be a fast run, but certainly possible.
By solving the problem first algebraically before plugging in any numbers, we are able to substitute other numbers as well, if we desire, without re-solving the whole problem.

P3.67. Prepare: We will use the initial information (that the marble goes 6.0 m straight up) to find the speed the marble leaves the gun. We also need to know how long it takes something to fall 1.5 m from rest in free fall so we can then use that in the horizontal equation.
Assume that there is no air resistance ($a_y = -g$) and that the marble leaves the gun with the same speed (muzzle speed) each time it is fired.
Solve: To determine the muzzle speed in the straight-up case, use Equation 2.13.

$$(v_y)_f^2 = (v_y)_i^2 + 2a_y \Delta y$$

where at the top of the trajectory $(v_y)_f = 0.0 \text{ m/s}$ and $\Delta y = 6.0 \text{ m}$.

$$(v_y)_i^2 = 2g\Delta y \Rightarrow (v_y)_i = \sqrt{2g\Delta y} = 10.8 \text{ m/s}$$

We also use Equation 2.14 to find the time for an object to fall 1.5 m from rest: $\Delta y = -15$ m now instead of the 6.0 m used previously.

$$\Delta y = \frac{1}{2}a_y(\Delta t)^2$$

$$\Delta t = \sqrt{\frac{2\Delta y}{-g}} = \sqrt{\frac{2(-1.5\text{ m})}{-9.8\text{ m/s}^2}} = 0.553\text{ s}$$

At last we combine this information into the equation for constant horizontal velocity.

$$\Delta x = v_x \Delta t = (10.8\text{ m/s})(0.553\text{ s}) = 6.0\text{ m}$$

Assess: Is it a coincidence that the marble has a horizontal range of 6.0 m when it can reach a height of 6.0 m when fired straight up, or will those numbers always be the same? Well, the 6.0 m horizontal range depends on the height (1.5 m) from which you fire it, so if that were different the range would be different. This leads us to conclude that it *is* a coincidence. You can go back, though, and do the problem algebraically (with no numbers) and find that g cancels and that the horizontal range is 2 times the square root of the product of the vertical height it can reach and the height from which you fire it horizontally.

P3.73. Prepare: First draw a picture.

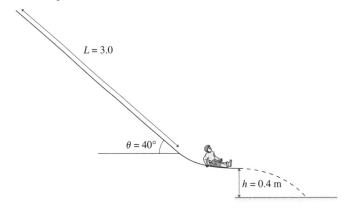

In part **(a)** use tilted axes so the *x*-axis runs down the slide. The acceleration will be $a_x = g\sin\theta$.

Part **(b)** is a familiar two-step projectile motion problem where we use the vertical direction to determine the time of flight and then plug it into then equation for constant horizontal velocity. Use axes that are *not* tilted for part **(b)**.

Solve: **(a)** We use Equation 2.13 with $(v_x)_i = 0.0$ m/s.

$$(v_x)_f^2 = 2a_x\Delta x$$

$$(v_x)_f = \sqrt{2(g\ \sin\theta)\Delta x} = \sqrt{2(9.8\text{ m/s}^2)(\sin 40°)(3.0\text{ m})} = 6.1\text{ m/s}$$

(b) We use Equation 2.14 to find the time for an object to fall 0.4 m from rest: $\Delta y = -0.4$m.

$$\Delta y = \frac{1}{2}a_y(\Delta t)^2$$

$$\Delta t = \sqrt{\frac{2\Delta y}{-g}} = \sqrt{\frac{2(-0.4\text{ m})}{-9.8\text{ m/s}^2}} = 0.286\text{ s}$$

At last we combine this information into the equation for constant horizontal velocity.

$$\Delta x = v_x\Delta t = (6.15\text{ m/s})(0.286\text{ s}) = 1.8\text{ m}$$

Assess: We reported the speed at the bottom of the slide to two significant figures, but kept track of a third to use as a guard digit because this result is also an intermediate result for the final answer. We also kept a third significant figure on the Δt as a guard digit.

The result of landing 1.8 m from the end of the frictionless slide seems just a bit large because this slide was frictionless and real slides aren't, but it doesn't seem to be too far out of expectation, so our result is probably correct.

P3.77. Prepare: We will use Equation 3.30 to relate the acceleration to the speed. But first we need to convert the speed of the car to m/s.

$$40\frac{\text{mi}}{\text{hr}} = 40\frac{\text{mi}}{\text{hr}}\left(\frac{1609\,\text{m}}{1\,\text{mi}}\right)\left(\frac{1\,\text{hr}}{3600\,\text{s}}\right) = 17.9 \text{ m/s}$$

Solve: (a) Your acceleration is given from the equation $a = v^2/r$

$$\frac{(17.9 \text{ m/s})^2}{110 \text{ m}} = 2.91 \text{ m/s}^2$$

which converts as follows:

$$2.91 \text{ m/s}^2 = \left(2.91 \text{ m/s}^2\right)\left(\frac{1\,g}{9.8 \text{ m/s}^2}\right) = 0.30\,g$$

The acceleration is 2.9 m/s^2 or $0.30\ g$.

(b) The formula for centripetal acceleration, $a = v^2/r$ can be solved for v as follows: $v = \sqrt{ar}$. In this form we see that if the acceleration is doubled, then the velocity is multiplied by $\sqrt{2}$. So we multiply the 40 mph speed limit by $\sqrt{2}$: $(40 \text{ mph})\sqrt{2} = 57 \text{ mph}$. At 57 mph the acceleration would be twice the acceleration at 40 mph.

Assess: As noted in the solution to Problem 40, a small change in velocity can produce a large change in centripetal acceleration. Here, with an increase in speed of less than 50%, the acceleration doubles and the friction needed for the turn also doubles.

Q4.1. Reason: If friction and air resistance are negligible (as stated) then the net force on the puck is zero (the normal force and gravitational force are equal in magnitude and opposite in direction). If the net force on the puck is zero, then Newton's first law states that it will continue on with constant velocity. So no force is needed to keep the puck moving; it will naturally keep moving unless a force acts on it to change its velocity.
Assess: This question demonstrates the difference between Aristotelian thinking and Newtonian thinking. Objects do not need forces on them to keep them moving; forces are only required when we want to *change* the velocity of the object. The reason one has to normally keep pushing an object to keep it moving is because of friction; to keep it at constant velocity your pushing force must be equal in magnitude to the friction force. But in the case of this question, there is no friction, so there is no force needed to keep it moving.

Q4.5. Reason: No. If you know all of the forces than you know the direction of the acceleration, not the direction of the motion (velocity). For example, a car moving forward could have on it a net force forward if speeding up or backward if slowing down or no net force at all if moving at constant speed.
Assess: Consider carefully what Newton's second law says, and what it doesn't say. The net force must *always* be in the direction of the acceleration. This is also the direction of the *change* in velocity, although not necessarily in the direction of the velocity itself.

Q4.7. Reason: The picture on the left is more effective at tightening the head because of the greater inertia of the head. Once moving, the head will "want" to continue moving (Newton's first law) after the handle hits the table, thus tightening the head, more so than in the second picture where the light handle has less inertia moving down than the head.
Assess: Newton's first law, the law of inertia, says the greater the mass of an object the more it will tend to continue with its previous velocity. One can assess this by trying it with a real hammer with a loose head.

Q4.9. Reason: Kinetic friction opposes the motion, but static friction is in the direction to prevent motion.
(a) Examples of motion where the frictional force on the object is directed opposite the motion would be a block sliding across a table top or the friction of the road on car tires in a skid.
(b) An example of motion in which the frictional force on the object is in the same direction as the motion would be a crate (not sliding) in the back of a pickup truck that is speeding up. The static frictional force of the truck bed on the crate is in the forward direction (the same direction as the motion) because it *is* the net force on the crate that accelerates it forward.
Assess: It is easy to think that the direction of the frictional force is *always* opposite the direction of motion, but static frictional forces are in the direction to prevent relative motion between the surfaces, and can be in various directions depending on the situation.

Q4.11. Reason: Both objects (Jonathan and his daughter) experience the same acceleration (i.e. the acceleration of the car). However since the objects do not have the same mass, the forces required to accelerate them will be different. The object with the greater mass (Jonathan) will require the greater force.
Assess: This is a straightforward application of Newton's second law.

Q4.13. Reason: The book defines weight as the gravitational pull of the earth on an object, and that doesn't change when the ball is thrown straight up. That is, if the gravitational force of the earth on the ball is 2.0 N while it sits on the scale, then the gravitational force of the earth on the ball is still 2.0 N while it is in flight—even at the very top of its motion. The weight of the ball is 2.0 N all the time the ball is near the surface of the earth, regardless of its motion.
Assess: Hmmm. . . . We previously distinguished between velocity, which *is* zero at the top of the trajectory (if the ball is thrown straight up), and acceleration, which is $-g$ all the while it is in free fall. But how do we explain

the fact that the acceleration of the ball is $-g$ at the top of its trajectory but zero while sitting on the scale, if the gravitational force on it is the same in both cases? Doesn't $\vec{F}_{net} = m\vec{a}$? Yes, $\vec{F}_{net} = m\vec{a}$ is true, but in the case of the ball on the scale the net force is zero because the scale is pushing up on the ball with a force of 2.0 N; since the net force is zero the acceleration is zero. While the ball is in flight the scale isn't pushing up on it, so the acceleration is $-g$ because the net force (the weight) is 2.0 N.

It is also worth mentioning that not all physics texts define weight the same way. Some define weight as "what the scale reads," which would be different from the definition in our text because if the scale and ball are thrown up together they would both be in free fall and the scale would read zero. That is, some books would say the ball's weight *is* zero at the top of its motion (because that is what the scale would read). Both definitions have merit, and it is wise to understand the difference and be prepared in future situations if someone defines weight differently than we have in this text. One way to avoid confusion is to always refer to "the gravitational force of the earth on the object" instead of the "weight." Still, the definition in our textbook is very mainstream.

Q4.19. Reason: The force that you exert on the wagon will cause it to move forward if it is greater than all opposing forces *on the wagon*. That is, the wagon will accelerate if the net force *on the wagon* is not zero. This is a proper application of Newton's second law, but you cannot apply Newton's third law in this case. Newton's third law does not apply to the forces on a single object, but only to forces acting on two different objects. A Newton's third law pair of forces can never cancel because they are always acting on different (opposite) objects.
Assess: It is nice to have smart three-year-olds, but in this case she needs even more physics understanding, not less. Both Newton's second law and his third law are true (in classical physics), but they don't address the same forces. Newton's second law addresses all of the forces acting on a single object; Newton's third law addresses pairs of forces that act on opposite objects (these third law forces can't even be added up—let alone cancel—because they aren't on the same object).

Problems

P4.1. Prepare: First note that time progresses to the right in each sequence of pictures. In one case the head is thrown back, and in the other, forward.
Solve: Using the principle of inertia, the head will tend to continue with the same velocity after the collision that it had before.
In the first series of sketches, the head is lagging behind because the car has been quickly accelerated forward (to the right). This is the result of a rear-end collision.
In the second series of sketches, the head is moving forward relative to the car because the car is slowing down and the head's inertia keeps it moving forward at the same velocity (although external forces do eventually stop the head as well). This is the result of a head-on collision.
Assess: Hopefully you haven't experienced either of these in an injurious way, but you have felt similar milder effects as the car simply speeds up or slows down.
It is for this reason that cars are equipped with headrests, to prevent the whiplash shown in the first series of sketches, because rear-end collisions are so common. The laws of physics tell us how wise it is to have the headrests properly positioned for our own height.
Air bags are now employed to prevent injury in the second scenario.

P4.5. Prepare: Draw the vector sum $\vec{F}_1 + \vec{F}_2$ of the two forces $\vec{F}_1$ and $\vec{F}_2$. Then look for a vector that will "balance" the force vector $\vec{F}_1 + \vec{F}_2$.

Solve: The object will be in equilibrium if $\vec{F}_3$ has the same magnitude as $\vec{F}_1 + \vec{F}_2$ but is in the opposite direction so that the sum of all three forces is zero.

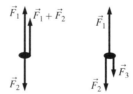

Assess: Adding the new force vector $\vec{F}_3$ with length and direction as shown will cause the object to be at rest.

P4.7. Prepare: Draw a picture of the situation, identify the system, in this case the mountain climber, and draw a closed curve around it. Name and label all relevant contact forces and long-range forces.

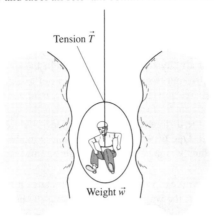

Solve: There are two forces acting on the mountain climber due to her interactions with the two agents earth and rope. One of the forces *on* the climber is the long-range weight force *by* the earth. The other force is the tension force exerted *by* the rope.

Assess: Note that the climber does not touch the sides of the crevasse so there are no forces from the crevasse walls.

P4.11. Prepare: We follow the outline in Tactics Box 4.2. See also Conceptual Example 4.2.

The exact angle of the slope is not critical in this problem; the answers would be very similar for any angle between $0°$ and $90°$.

Solve: The system is the skier.

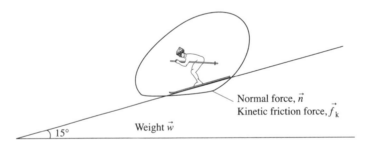

To identify forces, think of objects that are in contact with the object under consideration, as well as any long-range forces that might be acting on it. We are told to not ignore friction, but we will ignore air resistance.

The objects that are in contact with the skier are the snow-covered slope and. . . and that's all (although we will identify two forces exerted by this agent). The long-range force on the skier is the gravitational force of the earth on the skier.

One of the forces, then, is the gravitational force of the earth on the skier. This force points straight toward the center of the earth.

The slope, as we mentioned, exerts two forces on the skier: the normal force (directed perpendicularly to the slope) and the frictional force (directed parallel to the slope, backward from the downhill motion).

Assess: Since there are no other objects (agents) in contact with the skier (we are ignoring the air, remember?) and no other long-range forces we can identify (the gravitational force of the moon or the sun on the skier is also too small to be worth mentioning), then we have probably catalogued them all.

We are not told whether the skier has a constant velocity or is accelerating, and that factor would influence the relative lengths of the three arrows representing the forces. If the motion is constant velocity, then the vector sum of the three arrows must be zero.

P4.15. Prepare: Note that an object's acceleration is linearly proportional to the net force.

Solve: (a) One rubber band produces a force F, two rubber bands produce a force $2F$, and so on. Because $F \propto a$ and two rubber bands (force $2F$) produce an acceleration of 1.2 m/s^2, four rubber bands will produce an acceleration of 2.4 m/s^2.

(b) Now, we have two rubber bands (force 2F) pulling two glued objects (mass 2m). Using $F = ma, 2F = (2m)a \Rightarrow a = F/m = 0.60$ m/s^2.

Assess: Newton's second law predicts that for the same mass, doubling the force doubles the acceleration. It also says that doubling mass alone halves the acceleration. These are consistent with parts **(a)** and **(b)**, respectively.

P4.17. Prepare: The problem may be solved by applying Newton's second law to the present and the new situation.

Solve: **(a)** We are told that for an unknown force (call it F_o) acting on an unknown mass (call it m_o) the acceleration of the mass is 8.0 m/s^2. According to Newton's second law

$$F_o = m_o (8.0 \text{ m/s}^2) \quad \text{or} \quad F_o / m_o = 8.0 \text{ m/s}^2$$

For the new situation, the new force is $F_{new} = 2F_o$, the mass is not changed ($m_{new} = m_o$) and we may find the acceleration by

$$F_{new} = m_{new} a_{new}$$

or

$$a_{new} = F_{new} / m_{new} = 2F_o / m_o = 2(F_o / m_o) = 2(8 \text{ m/s}^2) = 16 \text{ m/s}^2$$

(b) For the new situation, the force is unchanged $F_{new} = F_o$, the new mass is half the old mass ($m_{new} = m_o / 2$) and we may find the acceleration by

$$F_{new} = m_{new} a_{new}$$

or

$$a_{new} = F_{new} / m_{new} = F_o / 2m_o = (F_o / m_o) / 2 = (8.0 \text{ m/s}^2) / 2 = 4.0 \text{ m/s}^2$$

(c) A similar procedure gives $a = 8.0$ m/s^2.
(d) A similar procedure gives $a = 32$ m/s^2.
Assess: From the algebraic relaionship $a = F / m$ we can see that when **(a)** the force is doubled, the acceleration is doubled; **(b)** the mass is doubled, the acceleration is halved; **(c)** both force and mass are doubled, the accelertaion doesn't change; and **(d)** force is doubled and mass is halved, the acceleration will be four times larger.

P4.21. Prepare: Refer to Figure P4.21.
Solve: Newton's second law is $F = ma$. We can read a force and an acceleration from the graph, and hence find the mass. Choosing the force $F = 1$ N gives us $a = 4$ m/s^2. Newton's second law then yields $m = 0.25$ kg.
Assess: Slope of the acceleration-versus-force graph is 4 m/N·s^2, and therefore, the inverse of the slope will give the mass.

P4.25. Prepare: This is a straightforward application of Newton's second law: $\vec{F}_{net} = m\vec{a}$, where $m = 3.0 \times 10^8$ kg, and $\vec{F}_{net} = \vec{F}_{thrust}$ (because we are told the supertanker is subject to no other forces), and $\vec{F}_{thrust} = 5 \times 10^6$ N (as given in Table 4.2).

The vectors $\vec{F}_{net}$ and $\vec{a}$ must point in the same direction (because m is never negative), so we now compute the magnitude.
Solve: Solve the second law for a.

$$a = \frac{F_{net}}{m} = \frac{F_{thrust}}{m} = \frac{5 \times 10^6 \text{ N}}{3.0 \times 10^8 \text{ kg}} = 0.0167 \text{ m/s}^2 \approx 0.02 \text{ m/s}^2$$

Assess: It appears that there is only one significant figure in the data in Table 4.2, hence the rounding to 0.02 m/s^2. If we assume one more significant figure in the table then we can report $a = 0.017$ m/s^2. However, the point of the problem is clear: The rocket motor (the largest force in Table 4.2) was able to give the supertanker only an impressively small acceleration.

P4.27. Prepare: The free-body diagram shows three forces with a net force (and therefore net acceleration) upward. There is a force labeled $\vec{w}$ directed down, a force $\vec{F}_{\text{thrust}}$ directed up, and a force $\vec{D}$ directed down. Now, draw a picture of a real object with three forces to match the given free-body diagram.

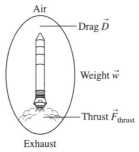

Solve: A possible description is, "A rocket accelerates upward."

Assess: It is given that the net force is pointing up. Then, $\vec{F}_{\text{net}} = \vec{F}_{\text{thrust}} - \vec{w} - \vec{D}$ must be greater than zero. In other words, $\vec{F}_{\text{thrust}}$ must be larger than $(\vec{w} + \vec{D})$.

P4.29. Prepare: Draw a picture of the situation, identify the system, in this case the car, and draw a closed curve around it. Name and label all relevant contact forces and long-range forces.

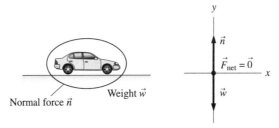

Solve: There are two forces acting *on* the car due to its interactions with the two agents the earth and the ground. One of the forces *on* the car is the long-range weight force *by* the earth. Another force is the normal force exerted *by* the ground due to the contact between the car and the ground. Since the car is sitting in the parking lot, acceleration is zero, and therefore the net force must also be zero. The free-body diagram is shown on the right.

Assess: It is implied that the car is sitting on a level parking lot, and therefore we need to consider only the vertical forces.

P4.31. Prepare: We follow the steps outlined in Tactics Boxes 4.2 and 4.3. We assume the road is level. We do not neglect air resistance, because at a "high speed" it is significant.

Solve: The system is your car.

(a) The objects in contact with your car are the air and the road. The road exerts two forces on the car: the normal force (directed up) and the kinetic friction force (directed horizontally back, parallel to the road). The air exerts a drag force (air resistance) on the car in the same direction as the friction force (i.e., opposite the velocity). The downward pull of the earth's gravitational force is the long-range force.

You could slow to a stop by air resistance alone if you are patient. You could also eventually slow to a stop by the friction of the road on the car (tires), but pressing on the brakes greatly increases the friction force and slows you down more quickly.

(b)

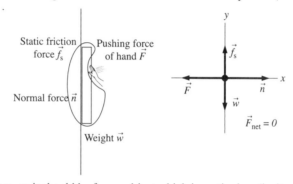

Normal force, $\vec{n}$
Kinetic friction force, $\vec{f}_k$
Air resistance, $\vec{D}$
Weight, $\vec{w}$

Assess: $\vec{F}_{net}$ points to the left, as does the $\vec{a}$ for a car that is moving to the right but slowing down.

P4.35. Prepare: We follow the steps outlined in Tactics Boxes 4.2 and 4.3.
Solve: The system is the picture.
The objects in contact with the picture are the wall and your hand. The wall exerts a normal force (opposite the pushing force of the hand) and a static friction force, which is directed upward and prevents the picture from falling down.
The important long-range force is the gravitational force of the earth on the picture (i.e., the weight).

Static friction force $\vec{f}_s$ Pushing force of hand $\vec{F}$

Normal force $\vec{n}$

Weight $\vec{w}$

$\vec{f}_s$

$\vec{F}$ $\vec{n}$

$\vec{w}$

$\vec{F}_{net} = 0$

Assess: The net force is zero, as it should be for an object which is motionless (isn't accelerating).

P4.37.   Prepare: Knowing that for every action there is an equal and opposite reaction and that these forces are exerted on different objects, we can identify, draw and label all the action-reaction pairs. Knowing all the forces acting on skater 2, we can construct a free-body diagram for skater 2.

Solve:

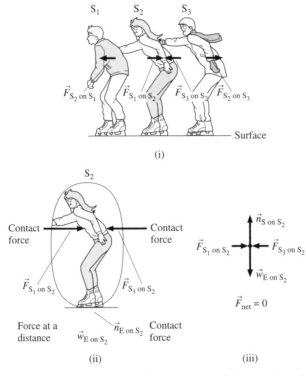

(i)

(ii)   (iii)

Figure (i) shows all three skater and the pair-wise interactions (connected by the dashed line) between them. Figure (ii) identifies skater 2 as the object of interest, shows the three contact forces that act on her, and the one long-range force that acts on her. Finally Figure (iii) shows a free body diagram for skater 2.

Assess: We have been informed that there is no friction and that the skaters are standing. If this is the case the net force acting on skater 2 should be zero. Also note that the action reaction pairs act on different objects.

P4.39. Prepare: Redraw the motion diagram as shown.

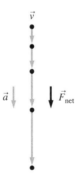

Solve: The previous figure shows velocity as downward, so the object is moving down. The length of the vector increases showing that the speed is increasing (like a dropped ball). Thus, the acceleration is directed down. Since $\vec{F} = m\vec{a}$, the force is in the same direction as the acceleration and must be directed down.

Assess: Since the object is speeding up, the acceleration vector must be parallel to the velocity vector and the net force must be parallel to the acceleration. In order to determine the net force, we had to combine our knowledge of motion diagrams, kinematics, and dynamics.

P4.43. Prepare: Redraw the motion diagram as shown.

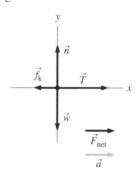

Solve: (a)–(c) Since the net force is to the right, the acceleration will also be to the right.
(d) There is a normal force and a weight, which are equal and opposite, so this is an object on a horizontal surface. The description could be, "A tow truck pulls a stuck car out of the mud."
Assess: Our scenario seems to fit the free body diagram. Check by doing the last part of the problem first: Start with the scenario and then draw a free-body diagram. Make sure it matches the original.

P4.45. Prepare: Redraw the motion diagram as shown.

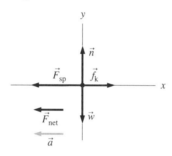

Solve: (a)–(c) The net force is to the left. Therefore acceleration will also be to the left.
(d) This is an object on a surface because $w = n$. It must be moving to the left because the kinetic friction is to the right. The description of the free-body diagram could be, "A compressed spring is shooting a plastic block to the left."
Assess: The scenario fits the free-body diagram. Check by doing the last part of the problem first: Start with the scenario and then draw a free-body diagram. Make sure it matches the original.

P4.51. Prepare: Review Tactics Box 4.3 about drawing free-body diagrams.

Solve: One error is that there isn't a force along the direction of motion in this case; $\vec{F}_{\text{motion}}$ should be erased completely.
Another error is that the drag force should be opposite the direction of the velocity, not straight left.
A correct free-body diagram would be

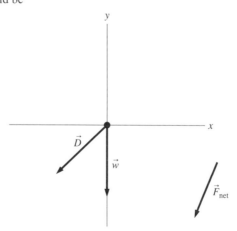

The acceleration (not shown) is in the same direction as $\vec{F}_{net}$. The velocity (not shown) is up to the right, opposite the drag force.

Assess: Motion is not a force. To draw a free-body diagram you must simply consider all of the forces acting *on* the object of interest. Do this by considering which objects are in contact with the object of interest, and which long-range forces act on the object.

P4.53. Prepare: There are three forces acting *on* the rocket due to its interactions with the three agents the earth, the air, and the hot gases exhausted to the environment. One force on the rocket is the long-range weight force *by* the earth. The second force is the drag force *by* the air. The third is the thrust force exerted on the rocket *by* the hot gas that is being let out to the environment. Since the rocket is being launched upward, it is being accelerated upward. Therefore, the net force on the rocket must also point upward. Draw a picture of the situation, identify the system, in this case the rocket, and draw a motion diagram. Draw a closed curve around the system, and name and label all relevant contact forces and long-range forces.

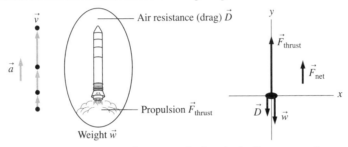

Solve: A force-identification diagram, a motion diagram, and a free-body diagram are shown.
Assess: You now have three important tools in your "Physics Toolbox," motion diagrams, force diagrams, and free-body diagrams. Careful use of these tools will give you an excellent conceptual understanding of a situation.

P4.55. Prepare: The normal force is perpendicular to the hill. The frictional force is parallel to the hill.

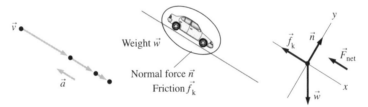

Solve: A force-identification diagram, a motion diagram, and a free-body diagram are shown.
Assess: You now have three important tools in your "Physics Toolbox," motion diagrams, force diagrams, and free-body diagrams. Careful use of these tools will give you an excellent conceptual understanding of a situation.

P4.59. Prepare: The ball rests on the floor of the barrel because the weight is equal to the normal force. There is a force of the spring to the right, which causes acceleration. The force of kinetic friction is smaller than the spring force. Now, draw a picture of the situation, identify the system, in this case the plastic ball, and draw a motion diagram. Draw a closed curve around the system, and name and label all relevant contact forces and long-range forces.

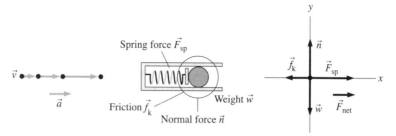

Solve: A force-identification diagram, a motion diagram, and a free-body diagram are shown.

Assess: Since the normal force acting on the ball and the weight of the ball are equal in magnitude and opposite in direction, the ball experiences no vertical motion. Since the spring exerts a greater force on the ball than kinetic friction, the ball accelerates out the barrel of the gun.

P4.61. Prepare: The gymnast experiences the long range force of weight. There is also a contact force from the trampoline, which is the normal force of the trampoline on the gymnast. The gymnast is moving downward and the trampoline is decreasing her speed, so the acceleration is upward and there is a net force upward. Thus the normal force must be larger than the weight. The actual behavior of the normal force will be complicated as it involves the stretching of the trampoline and therefore tensions.

Now, draw a picture of the situation, identify the system, in this case the gymnast, and draw a motion diagram. Draw a closed curve around the system, and name and label all relevant contact forces and long-range forces.

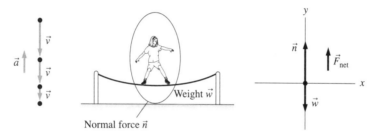

Solve: A force-identification diagram, a motion diagram, and a free-body diagram are shown.

Assess: There are only two forces on the gymnast. The weight force is directed downward and the normal force is directed upward. Since the gymnast is slowing down right after after making contact with the trampoline, upward normal force must be larger than the downward weight force.

P4.65. Prepare: To solve this problem, we will need to think carefully about the forces acting on the object of interest (the passenger in the car). It will also be important to determine what is meant by a *very* slippery bench.

Solve:

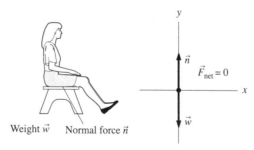

(a) The passenger is sitting on a *very* slippery bench in a car that is traveling to the right. Both the passenger and the seat are moving with a constant speed. There is a force on the passenger due to her weight, which is directed down. There is a contact force (the normal force) between the passenger and the seat, which is directed up. Since the passenger is not accelerating up or down, the net vertical force on her is zero, which means the two vertical forces are equal in magnitude. The statement of the problem gives no indication of any other contact forces. Specifically, we are told that the seat is *very* slippery. We can take this to mean there is no frictional force. So our force diagram includes only the normal force up, the weight down, but no horizontal force.

(b) The above considerations lead to the free-body diagram shown in the previous figure.

(c) The car (and therefore the *very slippery* seat) begins to slow down. Since the seat cannot exert a force of friction (either static or kinetic) on the passenger, the passenger cannot slow down as fast as the seat which is attached to the car. As a result the passenger continues to move forward with the initial speed of the car. Since the forces acting on the passenger remain the same, the free-body diagram is unchanged but the pictorial representation of the passenger is changed. These are shown in the following diagram.

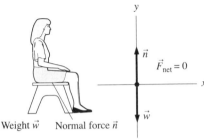

Weight $\vec{w}$ Normal force $\vec{n}$

(d) The car slows down because of some new contact force on the car (maybe the brakes lock the wheels and the road exerts a force on the tires). But there is no new contact force on you *the passenger*. The force diagram for the passenger remains unchanged, there are no horizontal forces on you. You do not slow down, but rather continue at an unchanged velocity until something in the picture changes for you (for example, you slide off the seat or hit the windshield). **(e)** The net force on *you* has remained zero because the net vertical force is zero and there are no horizontal forces. According to Newton's first law, if the net force on you is zero, then you continue to move in a straight line with a constant velocity. That is what happens to you when the car slows down. You continue to move forward with a constant velocity. The statement that you are "thrown forward" is misleading and incorrect. To be "thrown" there would need to be a net force on you and there is none. It might be correct to say that the car has been "thrown backward" leaving you to continue onward (until you part company with the seat).

Assess: Careful thinking and precise language, aided by a good diagram and understanding of Newton's first and second laws, are needed to articulate the solution of this problem.

P4.67. Prepare: The jump itself occurs while the froghopper is still in contact with the ground, but pushing off. During that time the froghopper must be accelerating upward, so the net force is upward.

Solve: **(a)**

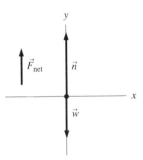

(b) Since the froghopper is accelerating upward the net force is upward; hence, the upward force of the ground on the froghopper is greater than the downward gravitational force (the froghopper's weight).

Assess: Once the froghopper leaves the ground it is accelerating down since it is slowing down; this is due to a net downward force (mainly the gravitational force). But before it leaves the ground it is accelerating upward.

APPLYING NEWTON'S LAWS

Q5.1. Reason: For an object to be in equilibrium, the net force (i.e., sum of the forces) must be zero. Assume that the two forces mentioned in the question are the only ones acting on the object.

The question boils down to asking if two forces can sum to zero if they aren't in opposite directions. Mental visualization shows that the answer is no, but so does a careful analysis. Set up a coordinate system with the x-axis along one of the forces. If the other force is not along the negative x-axis then there will be a y (or z) component that cannot be canceled by the first one along the x-axis.

Assess: In summary, two forces not in opposite directions cannot sum to zero. Neither can two forces with different magnitudes. However, three can.

Q5.5. Reason: The reading on the moon will be the moon-weight, or the gravitational force of the moon on the astronaut. This would be about 1/6 of the astronaut's earth-weight or the gravitational force of the earth on the astronaut (while standing on the scales on the earth).

Assess: Some physicists and textbooks define weight to be "what the scale reads" in which case it will read the astronaut's weight on the moon—by definition. But it won't read the same weight as on the earth. While not all physicists agree on the best definition of "weight" our textbook uses a very standard and reasonable approach. The astronaut's *mass* does not change by going to the moon.

Q5.9. Reason: The normal force (by definition) is directed perpendicular to the surface.
(a) If the surface that exerts a force on an object is vertical, then the normal force would be horizontal. An example would be holding a picture on a wall by pushing on it horizontally. The wall would exert a normal force horizontally.
(b) In a similar vein, if the surface that exerts a force on an object is horizontal and above the object, then the normal force would be down. One example would be holding a picture on a ceiling by pushing on it. The ceiling would exert a normal force vertically downward. Another example would be the Newton's third law pair force in the case of you sitting on a chair; the chair exerts a normal force upward on you, so you exert a normal force downward on the chair.

Assess: We see that the normal force can be in any direction; it is always perpendicular to the surface pushing on the object in question.

Q5.11. Reason: Use the simple model in Section 5.6 and assume that

$$D \approx \frac{1}{4} \rho A v^2$$

For object 1: $A = 0.20\,\text{m} \times 0.30\,\text{m} = 0.060\,\text{m}^2$; $v^2 = (6\,\text{m/s})^2 = 36\,\text{m}^2/\text{s}^2$; so $Av^2 = 2.2\,\text{m}^4/\text{s}^2$

For object 2: $A = 0.20\,\text{m} \times 0.20\,\text{m} = 0.040\,\text{m}^2$; $v^2 = (6\,\text{m/s})^2 = 36\,\text{m}^2/\text{s}^2$; so $Av^2 = 1.4\,\text{m}^4/\text{s}^2$

For object 3: $A = 0.30\,\text{m} \times 0.30\,\text{m} = 0.090\,\text{m}^2$; $v^2 = (4\,\text{m/s})^2 = 16\,\text{m}^2/\text{s}^2$; so $Av^2 = 1.4\,\text{m}^4/\text{s}^2$

The density of air ρ is the same for all three objects, so it won't affect the ranking.

Therefore, $D_1 > D_2 = D_3$.

Assess: Note that because v is squared, object 3's greater cross-sectional area did not produce the largest drag force.

Q5.15. Reason: If you only consider objects dropped from rest and accelerating up to terminal speed you might think that is the maximum speed the object can go through the air. However, terminal speed is merely the condition when the gravitational force and the air resistance force have the same magnitude and sum to zero. That doesn't necessarily mean it is the fastest possible speed.

It would be quite possible to throw or fire an object straight down from a high cliff at greater than terminal speed. The higher speed would mean that the upward drag force is greater than the downward gravitational force, so the net force would be up, the acceleration would be up, and the object would slow down to terminal speed, at which time the forces would cancel and the downward velocity would be constant.

Assess: The direction isn't crucial either; and object can also go up at faster than terminal speed, but the important aspects of the issue are most easily shown in the case above.

It is also possible to start from rest and speed up past terminal speed if there is another force which makes the net force non-zero and so acceleration can continue; an example of this would be a rocket-powered missile.

Q5.17. Reason: The tension is 49 N. It reads the same as it would if the rope were attached to the ceiling. The role of the five kilogram mass on the left is to keep the system in equilibrium, but it doesn't make the tension more than 49 N.

Assess: Apply Newton's second law to the mass on the right; the upward tension in the rope must equal the downward force of gravity. The scale reads the tension in the rope.

Problems

P5.1. Prepare: The massless ring is in static equilibrium, so all the forces acting on it must cancel to give a zero net force. The forces acting on the ring are shown on a free-body diagram below.

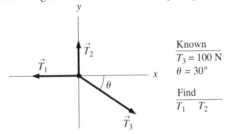

Solve: Written in component form, Newton's first law is

$$(F_{net})_x = \Sigma F_x = T_{1x} + T_{2x} + T_{3x} = 0 \text{ N} \qquad (F_{net})_y = \Sigma F_y = T_{1y} + T_{2y} + T_{3y} = 0 \text{ N}$$

Evaluating the components of the force vectors from the free-body diagram:

$$T_{1x} = -T_1 \qquad T_{2x} = 0 \text{ N} \qquad T_{3x} = T_3 \cos 30°$$

$$T_{1y} = 0 \text{ N} \qquad T_{2y} = T_2 \qquad T_{3y} = -T_3 \sin 30°$$

Using Newton's first law:

$$-T_1 + T_3 \cos 30° = 0 \text{ N} \qquad T_2 - T_3 \sin 30° = 0 \text{ N}$$

Rearranging:

$$T_1 = T_3 \cos 30° = (100 \text{ N})(0.8666) = 87 \text{ N} \qquad T_2 = T_3 \sin 30° = (100 \text{ N})(0.5) = 50 \text{ N}$$

Assess: Since $\vec{T}_3$ acts closer to the x-axis than to the y-axis, it makes sense that $T_1 > T_2$.

P5.3. Prepare: We assume the speaker is a particle in static equilibrium under the influence of three forces: gravity and the tensions in the two cables. So, all the forces acting on it must cancel to give a zero net force. The forces acting on the speaker are shown on a free-body diagram below. Because each cable makes an angle of 30° with the vertical, $\theta = 60°$.

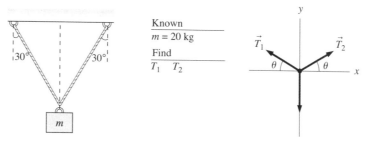

Solve: Newton's first law for this situation is

$$(F_{net})_x = \Sigma F_x = T_{1x} + T_{2x} = 0\,\text{N} \Rightarrow -T_1 \cos\theta + T_2 \cos\theta = 0\,\text{N}$$

$$(F_{net})_y = \Sigma F_y = T_{1y} + T_{2y} + w_y = 0\,\text{N} \Rightarrow T_1 \sin\theta + T_2 \sin\theta - w = 0\,\text{N}$$

The *x*-component equation means $T_1 = T_2$. From the *y*-component equation

$$2T_1 \sin\theta = w \Rightarrow T_1 = \frac{w}{2\sin\theta} = \frac{mg}{2\sin\theta} = \frac{(20\,\text{kg})(9.8\,\text{m/s}^2)}{2\sin 60°} = \frac{196\,\text{N}}{1.732} = 113\,\text{N}$$

Assess: It's to be expected that the two tensions are equal, since the speaker is suspended symmetrically from the two cables. This is 110 N to two significant figures.

P5.11. Prepare: Please refer to Figure P5.11. The free-body diagram shows three forces acting on an object whose mass is 2.0 kg. The force in the first quadrant has two components: 4 N along the *x*-axis and 3 N along the *y*-axis. We will first find the net force along the *x*- and the *y*-axes and then divide these forces by the object's mass to obtain the *x*- and *y*-components of the object's acceleration.
Solve: Applying Newton's second law to the diagram on the left:

$$a_x = \frac{(F_{net})_x}{m} = \frac{4\,\text{N} - 2\,\text{N}}{2\,\text{kg}} = 1.0\,\text{m/s}^2 \qquad a_y = \frac{(F_{net})_y}{m} = \frac{3\,\text{N} - 3\,\text{N}}{2\,\text{kg}} = 0\,\text{m/s}^2$$

Assess: The object's motion is only along the *x*-axis.

P5.13. Prepare: We assume that the box is a particle being pulled in a straight line. Since the ice is frictionless, the tension in the rope is the only horizontal force on the box and is shown below in the free-body diagram. Since we are looking at horizontal motion of the box, we are not interested in the vertical forces in this problem.

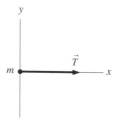

Solve: **(a)** Since the box is at rest, $a_x = 0$ m/s², the net force on the box must be zero or the tension in the rope must be zero.

(b) For this situation again, $a_x = 0$ m/s², so $F_{net} = T = 0$ N.

(c) Here, the velocity of the box is irrelevant, since only a *change* in velocity requires a nonzero net force. Since $a_x = 5.0$ m/s²,

$$F_{net} = T = ma_x = (50\,\text{kg})(5.0\,\text{m/s}^2) = 250\,\text{N}$$

Assess: For parts **(a)** and **(b)**, the zero acceleration immediately implies that the rope is exerting no horizontal force on the box. For part **(c)**, the 250 N force (the equivalent of about half the weight of a small person) seems reasonable to accelerate a box of this mass at 5.0 m/s².

P5.15. Prepare: We assume that the seat belt supplies all the force necessary to decelerate the driver (that is, $F_{\text{seatbelt}} = F_{\text{net}}$), and that the deceleration is constant over the time interval of 0.10 s. Set up a coordinate system with the car traveling to the right along the *x*-axis.

We use the kinematics equations from Chapter 2 to solve for the constant acceleration, and then $F_{\text{net}} = ma$ (with $m = 70$ kg) to solve for the force exerted by the seat belt.

Solve: The definition of acceleration says

$$a_x = \frac{\Delta v_x}{\Delta t} = \frac{0.0 \text{ m/s} - 14 \text{ m/s}}{0.10 \text{ s}} = -140 \text{ m/s}^2$$

where the negative sign indicates that the car (which is traveling to the right) is slowing down.

$$F_{\text{seatbelt}} = F_{\text{net}} = ma_x = (70 \text{ kg})(-140 \text{ m/s}^2) = -9800 \text{ N}$$

where the negative sign shows the force acting in the negative *x*-direction (the same direction as the acceleration).

Assess: 9800 N is quite a bit of force, but so it is in a head-on collision at a significant speed. You can see from the equations above that if the crash had taken more time the force would not be so severe; save that thought for a future chapter.

P5.17. Solve: (a) The woman's weight on the earth is

$$w_{\text{earth}} = mg_{\text{earth}} = (55 \text{ kg})(9.8 \text{ m/s}^2) = 540 \text{ N}$$

(b) Since mass is a measure of the amount of matter, the woman's mass is the same on the moon as on the earth. Her weight on the moon is

$$w_{\text{moon}} = mg_{\text{moon}} = (55 \text{ kg})(1.62 \text{ m/s}^2) = 89 \text{ N}$$

Assess: The smaller acceleration due to gravity on the moon reveals that objects are less strongly attracted to the moon than to the earth. Thus the woman's smaller weight on the moon makes sense.

P5.19. Prepare: The astronaut and the chair will be denoted by A and C, respectively, and they are separate systems. The launch pad is a part of the environment. In the following free-body diagrams for both the astronaut and the chair are shown at rest on the launch pad (top) and while accelerating (bottom).

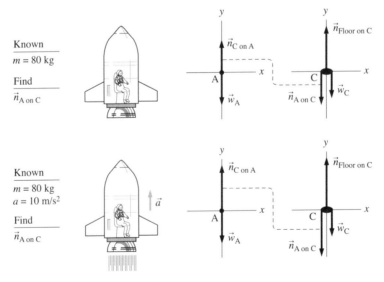

Solve: (a) Newton's second law for the astronaut is

$$\Sigma (F_{\text{on A}})_y = n_{\text{C on A}} - w_A = m_A a_A = 0 \text{ N} \Rightarrow n_{\text{C on A}} = w_A = m_A g$$

By Newton's third law, the astronaut's force on the chair is

$$n_{\text{A on C}} = n_{\text{C on A}} = m_A g = (80 \text{ kg})(9.8 \text{ m/s}^2) = 780 \text{ N}$$

(b) Newton's second law for the astronaut is

$$\Sigma(F_{\text{on A}})_y = n_{\text{C on A}} - w_A = m_A a_A \Rightarrow n_{\text{C on A}} = w_A + m_A a_A = m_A(g + a_A)$$

By Newton's third law, the astronaut's force on the chair is

$$n_{\text{A on C}} = n_{\text{C on A}} = m_A(g + a_A) = (80 \text{ kg})(9.8 \text{ m/s}^2 + 10 \text{ m/s}^2) = 1600 \text{ N}$$

Assess: This is a reasonable value because the astronaut's acceleration is more than *g*.

P5.23. Prepare: In each case the frog is in equilibrium ($\vec{F}_{\text{net}} = \vec{0}$).

Solve: **(a)** The two forces on the frog act in the vertical direction: the weight (gravitational force of the earth down on the frog), and the normal force of the log up on the frog. The two must have equal magnitude; since $w = mg$ (0.60 kg) (9.8 m/s^2) = 5.9 N, then the magnitude of the normal force is also 5.9 N.
(b) Draw a free-body diagram for the frog. Use tilted axes with the *x*-axis running up the log.

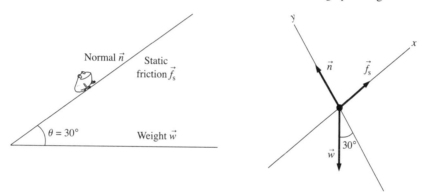

Apply $F_{\text{net}} = ma$ in the *y*-direction.

$$n - w \cos\theta = 0$$
$$n = w \cos\theta = mg \cos\theta = (0.60 \text{ kg})(9.8 \text{ m/s}^2)\cos 30° = 5.1 \text{ N}$$

Assess: The answer is less in part **(b)** than in part **(a)**, as we would expect. The static friction force is also helping hold up the frog in part **(b)**.
Notice that we solved the problem algebraically before putting numbers in. This not only allows us to solve a similar problem for a different frog or log, but it enables us to check our answer in this case for reasonableness. Take the limit as $\theta \to 0$; the slope approaches zero and the conditions revert back to part **(a)** as $\cos\theta \to 1$. Then take the limit as $\theta \to 90°$ and the normal force decreases to zero as the log becomes vertical and there is no normal force on the frog.

P5.25. Prepare: We assume that the safe is a particle moving only in the *x*-direction. Since it is sliding during the entire problem, the force of kinetic friction opposes the motion by pointing to the left. In the following diagram we give a pictorial representation and a free-body diagram for the safe. The safe is in dynamic equilibrium, since it's not accelerating.

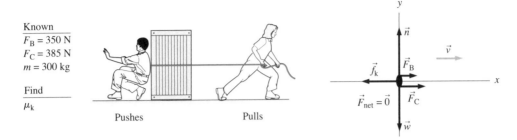

Known
$F_B = 350$ N
$F_C = 385$ N
$m = 300$ kg

Find
μ_k

Pushes   Pulls

Solve: We apply Newton's first law in the vertical and horizontal directions:

$$(F_{net})_x = \Sigma F_x = F_B + F_C - f_k = 0 \text{ N} \Rightarrow f_k = F_B + F_C = 350 \text{ N} + 385 \text{ N} = 735 \text{ N}$$

$$(F_{net})_y = \Sigma F_y = n - w = 0 \text{ N} \Rightarrow n = w = mg = (300 \text{ kg})(9.8 \text{ m/s}^2) = 2940 \text{ N}$$

Then, for kinetic friction

$$f_k = \mu_k n \Rightarrow \mu_k = \frac{f_k}{n} = \frac{735 \text{ N}}{2940 \text{ N}} = 0.25$$

Assess: The value of $\mu_k = 0.25$ is hard to evaluate without knowing the material the floor is made of, but it seems reasonable.

P5.27. Prepare: The car is undergoing skidding, so it is decelerating and the force of kinetic friction acts to the left. We give below an overview of the pictorial representation, a motion diagram, a free-body diagram, and a list of values. We will first apply Newton's second law to find the deceleration and then use kinematics to obtain the length of the skid marks.

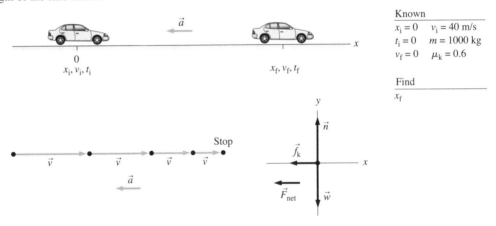

Known
$x_i = 0$ $v_i = 40$ m/s
$t_i = 0$ $m = 1000$ kg
$v_f = 0$ $\mu_k = 0.6$

Find
x_f

Solve: We begin with Newton's second law. Although the motion is one-dimensional, we need to consider forces in both the x- and y-directions. However, we know that $a_y = 0 \text{ m/s}^2$. We have

$$a_x = \frac{(F_{net})_x}{m} = \frac{-f_k}{m} \qquad a_y = 0 \text{ m/s}^2 = \frac{(F_{net})_y}{m} = \frac{n - w}{m} = \frac{n - mg}{m}$$

We used $(f_k)_x = -f_k$ because the free-body diagram tells us that $\vec{f}_k$ points to the left. The force of kinetic friction relates $\vec{f}_k$ to $\vec{n}$ with the equation $f_k = \mu_k n$. The y-equation is solved to give $n = mg$. Thus, the kinetic friction force is $f_k = \mu_k mg$.

Substituting this into the x-equation yields

$$a_x = \frac{-\mu_k mg}{m} = -\mu_k g = -(0.6)(9.8 \text{ m/s}^2) = -5.88 \text{ m/s}^2$$

The acceleration is negative because the acceleration vector points to the left as the car slows. Now we have a constant-acceleration kinematics problem. Δt isn't known, so use

$$v_f^2 = 0 \text{ m}^2/\text{s}^2 = v_i^2 + 2a_x \Delta x \Rightarrow \Delta x = -\frac{(40 \text{ m/s})^2}{2(-5.88 \text{ m/s}^2)} = 140 \text{ m}$$

Assess: The skid marks are 140 m long. This is ≈ 430 feet, reasonable for a car traveling at ≈ 80 mph. It is worth noting that an algebraic solution led to the m canceling out.

P5.33. Prepare: We assume that the skydiver is shaped like a box. The following shows a pictorial representation of the skydiver and a free-body diagram at terminal speed. The skydiver falls straight down toward the earth's surface, that is, the direction of fall is vertical. Since the skydiver falls feet first, the surface perpendicular to the drag has the cross-sectional area $A = 20 \text{ cm} \times 40 \text{ cm}$. The physical conditions needed to use Equation 5.15 for the drag force to be satisfied. The terminal speed corresponds to the situation when the net force acting on the skydiver becomes zero.

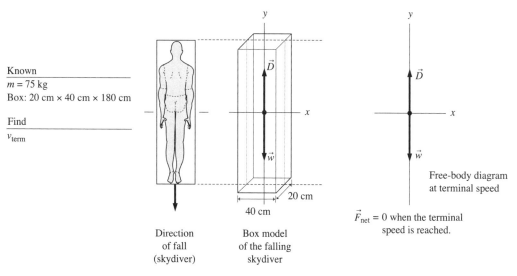

Solve: The expression for the magnitude of the drag with v in m/s is

$$D \approx \frac{1}{4}\rho A v^2 = 0.25(1.22 \text{ kg/m}^3)(0.20 \text{ m} \times 0.40 \text{ m})v^2 \text{ N} = 0.0244v^2 \text{ N}$$

The skydiver's weight is $w = mg = (75 \text{ kg})(9.8 \text{ m/s}^2) = 735 \text{ N}$. The mathematical form of the condition defining dynamical equilibrium for the skydiver and the terminal speed is

$$\vec{F}_{net} = \vec{w} + \vec{D} = 0 \text{ N} \Rightarrow 0.0244v_{term}^2 \text{ N} - 735 \text{ N} = 0 \text{ N} \Rightarrow v_{term} = \sqrt{\frac{735}{0.0244}} \approx 170 \text{ m/s}$$

Assess: The result of the above simplified physical modeling approach and subsequent calculation, even if approximate, shows that the terminal velocity is very high. This result implies that the skydiver will be very badly hurt at landing if the parachute does not open in time.

P5.35. Prepare: The blocks are denoted as 1, 2, and 3. The surface is frictionless and along with the earth it is a part of the environment. The three blocks are our three systems of interest. The force applied on block 1 is $F_{A\,on\,1} = 12$ N. The acceleration for all the blocks is the same and is denoted by a. A visual overview shows a pictorial representation, a list of known and unknown values, and a free-body diagram for the three blocks.

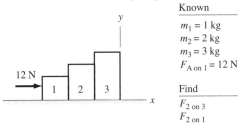

Known
$m_1 = 1$ kg
$m_2 = 2$ kg
$m_3 = 3$ kg
$F_{A\,on\,1} = 12$ N

Find
$F_{2\,on\,3}$
$F_{2\,on\,1}$

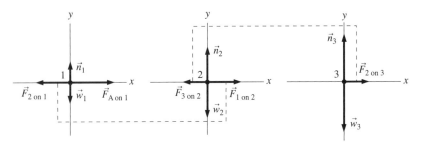

Solve: Newton's second law for the three blocks along the x-direction is

$$\Sigma(F_{on\,1})_x = F_{A\,on\,1} - F_{2\,on\,1} = m_1 a \qquad \Sigma(F_{on\,2})_x = F_{1\,on\,2} - F_{3\,on\,2} = m_2 a \qquad \Sigma(F_{on\,3})_x = F_{2\,on\,3} = m_3 a$$

Adding these three equations and using Newton's third law ($F_{2\,on\,1} = F_{1\,on\,2}$ and $F_{3\,on\,2} = F_{2\,on\,3}$), we get

$$F_{A\,on\,1} = (m_1 + m_2 + m_3)a \Rightarrow (12\text{ N}) = (1\text{ kg} + 2\text{ kg} + 3\text{ kg})a \Rightarrow a = 2\text{ m/s}^2$$

Using this value of a, the force equation on block 3 gives

$$F_{2\,on\,3} = m_3 a = (3\text{ kg})(2\text{ m/s}^2) = 6\text{ N}$$

Substituting into the force equation on block 1,

$$12\text{ N} - F_{2\,on\,1} = (1\text{ kg})(2\text{ m/s}^2) \Rightarrow F_{2\,on\,1} = 10\text{ N}$$

Assess: Because all three blocks are pushed forward by a force of 12 N, the value of 10 N for the force that the 2 kg block exerts on the 1 kg block is reasonable.

P5.39. Prepare: Since each block has the same acceleration as all the others they must each experience the same net force. Each block will have one more newton pulling forward than the force pulling back on it from the blocks behind.
Solve:
(a) 1 N
(b) 50 N
Assess: Since 100 N accelerates 100 blocks then n newtons accelerates n blocks.

P5.41. Prepare: Because the piano is to descend at a steady speed, it is in dynamic equilibrium. The following shows a free-body diagram of the piano and a list of values.

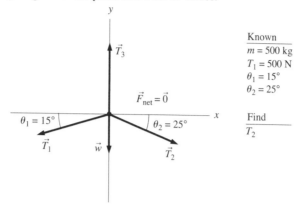

Known
$m = 500$ kg
$T_1 = 500$ N
$\theta_1 = 15°$
$\theta_2 = 25°$

Find

T_2

Solve: **(a)** Based on the free-body diagram, Newton's second law is

$$(F_{net})_x = 0 \text{ N} = T_{1x} + T_{2x} = T_2 \cos\theta - T_1 \cos\theta_1$$
$$(F_{net})_y = 0 \text{ N} = T_{1y} + T_{2y} + T_{3y} + w_y = T_3 - T_1 \sin\theta_1 - T_2 \sin\theta_2 - mg$$

Notice how the force components all appear in the second law with *plus* signs because we are *adding* vector forces. The negative signs appear only when we *evaluate* the various components. These are two simultaneous equations in the two unknowns T_2 and T_3. From the *x*-equation we find

$$T_2 = \frac{T_1 \cos\theta_1}{\cos\theta_2} = \frac{(500 \text{ N})\cos 15°}{\cos 25°} = 530 \text{ N}$$

(b) Now we can use the *y*-equation to find

$$T_3 = T_1 \sin\theta_1 + T_2 \sin\theta_2 + mg = 5300 \text{ N}$$

P5.49. Prepare: The rock (R) and Bob (B) are two systems of our interest. We give a pictorial representation, a list of values, and free-body diagrams for Bob and the rock. Motion of the rock is assumed to be along the *x*-axis. We realize that Bob must accelerate the rock forward until he releases the rock with a speed of 30 m/s. From the given information we can find this acceleration using kinematics. Newton's second law will then yield the force exerted on the rock by Bob. This is also the force that is exerted by the rock on Bob (Newton's third law). We can then calculate Bob's acceleration using his mass in Newton's second law. Kinematics once again can be used to find Bob's recoil speed.

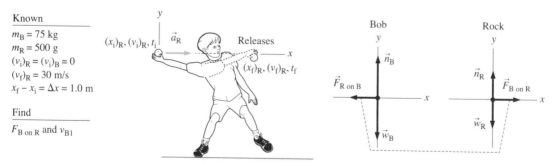

Known
$m_B = 75$ kg
$m_R = 500$ g
$(v_i)_R = (v_i)_B = 0$
$(v_f)_R = 30$ m/s
$x_f - x_i = \Delta x = 1.0$ m

Find

$F_{B \text{ on } R}$ and v_{B1}

Solve: **(a)** Bob exerts a forward force $\vec{F}_{B \text{ on } R}$ on the rock to accelerate it forward. The rock's acceleration is calculated as follows:

$$(v_f)_R^2 = (v_i)_R^2 + 2a_R \Delta x \Rightarrow a_R = \frac{(v_f)_R^2}{2\Delta x} = \frac{(30 \text{ m/s})^2}{2(1 \text{ m})} = 450 \text{ m/s}^2$$

The force is calculated from Newton's second law:

$$F_{B\text{ on }R} = m_R a_R = (0.5\text{ kg})(450\text{ m/s}^2) = 225\text{ N or }230\text{ N to two significant figures.}$$

(b) Because Bob pushes on the rock, the rock pushes back on Bob with a force $\vec{F}_{R\text{ on }B}$. Forces $\vec{F}_{R\text{ on }B}$ and $\vec{F}_{R\text{ on }B}$ are an action/reaction pair, so $\vec{F}_{R\text{ on }B} = \vec{F}_{B\text{ on }R} = 225\text{ N}$. The force causes Bob to accelerate backward with an acceleration equal to

$$a_B = \frac{(F_{\text{net on }B})_x}{m_B} = -\frac{F_{R\text{ on }B}}{m_B} = -\frac{225\text{ N}}{75\text{ kg}} = -3.0\text{ m/s}^2$$

This is a rather large acceleration, but it lasts only until Bob releases the rock. We can determine the time interval by returning to the kinematics of the rock:

$$(v_f)_R = (v_i)_R + a_R \Delta t \Rightarrow \Delta t = \frac{(v_f)_R}{a_R} = 0.0667\text{ s}$$

At the end of this interval, Bob's velocity is

$$(v_f)_B = (v_i)_B + a_B \Delta t = a_B \Delta t = -0.20\text{ m/s}$$

Thus his recoil speed is 0.20 m/s.

P5.51. Prepare: We assume the rocket is moving in a vertical straight line along the *y*-axis under the influence of only two forces: gravity and its own thrust. The free-body diagram for the model rocket is shown later.

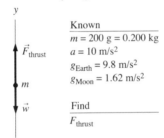

Known
$m = 200\text{ g} = 0.200\text{ kg}$
$a = 10\text{ m/s}^2$
$g_{\text{Earth}} = 9.8\text{ m/s}^2$
$g_{\text{Moon}} = 1.62\text{ m/s}^2$

Find
F_{thrust}

Solve: **(a)** Using Newton's second law and reading the forces from the free-body diagram,

$$F_{\text{thrust}} - w = ma \Rightarrow F_{\text{thrust}} = ma + mg_{\text{earth}} = (0.200\text{ kg})(10\text{ m/s}^2 + 9.8\text{ m/s}^2) = 4.0\text{ N}$$

(b) Likewise, the thrust on the moon is $(0.200\text{ kg})(10\text{ m/s}^2 + 1.62\text{ m/s}^2) = 2.3\text{ N}$.
Assess: The thrust required is smaller on the moon, as it should be, given the moon's weaker gravitational pull. The magnitude of a few newtons seems reasonable for a small model rocket.

P5.55. Prepare: The length of the hill is $\Delta x = h / \sin\theta$. The acceleration is $g\sin\theta$.
Solve: First use the kinematic equation, with $v_i = 0\text{ m/s}$ at the top of the hill, to determine the speed at the bottom of the hill.

$$(v_f)_1^2 = (v_i)_1^2 + 2a\Delta x \quad \Rightarrow \quad (v_f)_1^2 = 2(g\sin\theta)(h/\sin\theta) = 2gh$$

Now apply the same kinematic equation to the horizontal patch of snow, only this time we want Δx. To connect the two parts $(v_f)_1 = (v_i)_2$. The final speed is zero: $(v_f)_2 = 0$.

$$(v_f)_2^2 = (v_i)_2^2 + 2a\Delta x = (v_f)_1^2 + 2a\Delta x = 2gh + 2a\Delta x = 0$$

The friction force is the net force, so $a = -f_k / m$. Note $f_k = \mu_k n = \mu_k mg$. Solve for Δx.

$$\Delta x = \frac{-2gh}{2a} = \frac{-gh}{-f_k / m} = \frac{gh}{\mu_k mg / m} = \frac{h}{\mu_k} = \frac{3.0\text{ m}}{0.05} = 60\text{ m}$$

Assess: It seems reasonable to glide 60 m with such a low coefficient of friciton. It is interesting that we did not need to know the angle of the (frictionless) slope; this will become clear in the chapter on energy. The answer is also independent of Josh's mass.

P5.59.   Prepare: Sam is moving along the *x*-axis under the influence of two forces: the thrust of his jet skis and the resisting force of kinetic friction on the skis. A visual overview of Sam's motion is shown below in a pictorial representation, motion diagram, free-body diagrams in the accelerating and coasting periods, and a list of values. To find Sam's top speed using kinematics, we will first find his acceleration from Newton's second law. Kinematic equations will then allow us to find his displacement during both accelerating and coasting intervals.

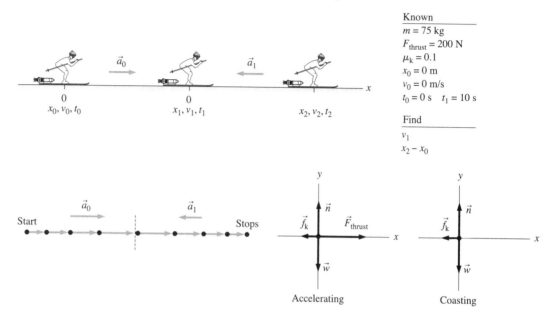

Solve: (a) The friction force of the snow can be found from the free-body diagram and Newton's first law, since there's no acceleration in the vertical direction:

$$n = w = mg = (75\,\text{kg})(9.8\,\text{m/s}^2) = 735\,\text{N} \Rightarrow f_k = \mu_k n = (0.1)(735\,\text{N}) = 73.5\,\text{N}$$

Then, from Newton's second law,

$$(F_{\text{net}})_x = F_{\text{thrust}} - f_k = ma_0 \Rightarrow a_0 = \frac{F_{\text{thrust}} - f_k}{m} = \frac{200\,\text{N} - 73.5\,\text{N}}{75\,\text{kg}} = 1.687\,\text{m/s}^2$$

From kinematics:

$$v_1 = v_0 + a_0 t_1 = 0\,\text{m/s} + (1.687\,\text{m/s}^2)(10\,\text{s}) = 16.87\,\text{m/s or } 17\,\text{m/s to two significant figures.}$$

(b) During the acceleration, Sam travels to

$$x_1 = x_0 + v_0 t_1 + \frac{1}{2} a_0 t_1^2 = \frac{1}{2}(1.687\,\text{m/s}^2)(10\,\text{s})^2 = 84\,\text{m}$$

After the skis run out of fuel, Sam's acceleration can again be found from Newton's second law:

$$(F_{\text{net}})_x = -f_k = -73.5\,\text{N} \Rightarrow a_1 = \frac{F_{\text{net}}}{m} = \frac{-73.5\,\text{N}}{75\,\text{kg}} = -0.98\,\text{m/s}^2$$

Since we don't know how much time it takes Sam to stop,

$$v_2^2 = v_1^2 + 2a_1(x_2 - x_1) \Rightarrow x_2 - x_1 = \frac{v_2^2 - v_1^2}{2a_1} = \frac{0\,\text{m}^2/\text{s}^2 - (16.87\,\text{m/s})^2}{2(-0.98\,\text{m/s}^2)} = 145\,\text{m}$$

The total distance traveled is $(x_2 - x_1) + x_1 = 145\,\text{m} + 84\,\text{m} = 229\,\text{m or } 230\,\text{m to two significant figures.}$

Assess: A top speed of 16.9 m/s (roughly 40 mph) seems quite reasonable for this acceleration, and a coasting distance of nearly 150 m also seems possible, starting from a high speed, given that we're neglecting air resistance.

P5.61. Prepare: We show below the free-body diagram of the 1 kg block. The block is initially at rest, so initially the friction force is static friction. If the 12 N pushing force is too strong, the box will begin to move up the wall. If it is too weak, the box will begin to slide down the wall. And if the pushing force is within the proper range, the box will remain stuck in place.

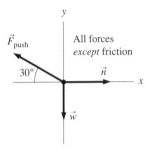

Solve: First, let's evaluate the sum of all the forces *except* friction:

$$\Sigma F_x = n - F_{push} \cos 30° = 0 \text{ N} \Rightarrow n = F_{push} \cos 30°$$

$$\Sigma F_y = F_{push} \sin 30° - w = F_{push} \sin 30° - mg = (12 \text{ N})\sin 30° - (1 \text{ kg})(9.8 \text{ m/s}^2) = -3.8 \text{ N}$$

In the first equation we have utilized the fact that any motion is parallel to the wall, so $a_x = 0 \text{ m/s}^2$.
The two forces in the second y-equation add up to -3.8 N. This means the static friction force will be able to prevent the box from moving if $f_s = +3.8 \text{ N}$. Using the x-equation we get

$$f_{s \text{ max}} = \mu_s n = \mu_s F_{push} \cos 30° = 5.2 \text{ N}$$

where we used $\mu_s = 0.5$ for wood on wood. The static friction force $\vec{f}_s$ needed to keep the box from moving is *less* than $f_{s \text{ max}}$. Thus the box will stay at rest.

P5.67. Prepare: Since the block comes to rest for an instant, we use the coefficient of static friction for wood on wood: $\mu_s = 0.50$.
Solve:

$$\Sigma F_x = mg \sin \theta - \mu_s n = 0$$

$$\Sigma F_y = n - mg \cos \theta = 0$$

Solve the second equation for n and insert into the first.

$$mg \sin \theta - \mu_s mg \cos \theta = 0$$

$$\tan \theta = \mu_s \quad \Rightarrow \quad \theta = \tan^{-1}(\mu_s) = \tan^{-1}(0.50) = 26.57° \approx 27°$$

Assess: From experience, $27°$ seems like a reasonable tilt for the block to slide back down.

P5.69. Prepare: The Ping-Pong ball when shot straight up is subject to a net force that is the resultant of the weight and drag force vectors, both acting vertically downward. On the other hand, for the ball's motion straight down, the ball is subject to a net force that is the resultant of the weight and drag force vectors, the former in the downward and the latter in the upward direction. An overview of a pictorial representation and a free-body diagram are shown below. The Ping-Pong ball experiences a drag force equal to $\frac{1}{4}\rho A v^2$, as modeled in the text with v_{term} as the terminal velocity.

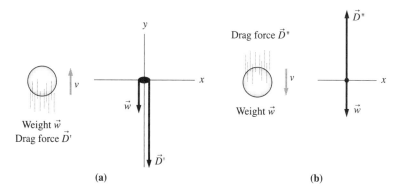

(a) **(b)**

Solve: **(a)** Imagine the ball falling at its terminal speed. The ball's weight is directed down and the resistive drag force is directed up. The net force is zero because the magnitude of the drag force is equal to the magnitude of the weight, $D = w$. When the ball is shot upward at twice the terminal speed, the drag force is four times the terminal drag force. That is, $D' = 4D = 4w$.

Since all the forces are down, the *y*-component of Newton's second law is

$$\sum F_y = -D' - w = -4w - w = -5mg = ma \Rightarrow a = -5g$$

(b) The ball is initially shot downward. Therefore D'' is upward but *w* is down. Again $D'' = 4D$ and the *y*-component of Newton's second law is

$$\sum F_y = D'' - w = 4w - w = 3mg = ma \Rightarrow a = 3g$$

That is, the ball initially decelerates at $3g$ but as *v* becomes smaller, the drag force approaches the weight so the deceleration goes to zero and *v* approaches v_{term}.

Assess: D' is very large and with *w* yields a large initial deceleration when the ball is shot up. When the ball is shot down *w* opposes D'' so the ball decelerates at a lesser rate.

P5.71. Prepare: Call the 10 kg block m_2 and the 5.0 kg block m_1. Assume the pulley is massless and frictionless.

Solve: On block 2 use tilted axes.

$$\sum F_x = T - = m_2 g \sin \theta = m_2 a_2$$

Block 1 is also accelerating.

$$\sum F_y = T - m_1 g = m_1 a_1$$

The acceleration constraint is $(a_2)_x = -(a_1)_y = a$. Solve for *T* in the second equation and insert in the first. $T = m_1(g - a)$.

$$m_1(g - a) - m_2 g \sin \theta = m_2 a$$

$$m_1 g - m_2 g \sin \theta = m_2 a + m_1 a$$

$$a = \frac{g(m_1 - m_2 \sin \theta)}{m_1 + m_2} = \frac{(9.8 \text{ m/s}^2)(1.0 \text{ kg} - (2.0 \text{ kg}) \sin 40°)}{1.0 \text{ kg} + 2.0 \text{ kg}} = 1.96 \text{ m/s}^2 \approx 2.0 \text{ m/s}^2 = -0.93 \text{ m/s}^2$$

Or 0.93 m/s^2, down the ramp.

Assess: The answer depends on θ; for a shallow angle the block accelerates up the ramp, for a steep angle the block accelerates down the ramp. This is expected behavior.

P5.73. Prepare: We assume the dishes have a constant acceleration during the 0.25 s, and we'll use a kinematic equation as well as Newton's second law. The friction force is the net force, $F_{net} = f_k = \mu_k mg$. We also know $a = F_{net} / m$.

Solve: Use the kinematic equation (where $v_i = 0$):

$$\Delta x = \frac{1}{2}a(\Delta t)^2 = \frac{1}{2}\frac{F_{net}}{m}(\Delta t)^2 = \frac{1}{2}\frac{\mu_k mg}{m}(\Delta t)^2 = \frac{1}{2}\mu_k g(\Delta t)^2 = \frac{1}{2}(0.12)(9.8 \text{ m/s}^2)(0.25 \text{ s})^2 = 3.7 \text{ cm}$$

Assess: This seems to be a reasonable distance the dishes could travel without falling off the edge.

P5.77. Solve: **(a)** A driver traveling at 40 m/s in her 1500 kg auto slams on the brakes and skids to rest. How far does the auto slide before coming to rest?
(b)

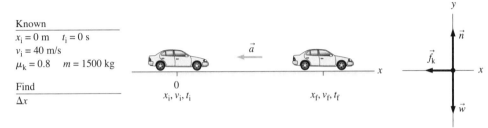

(c) Newton's second law is

$$\sum F_y = n_y + w_y = n - mg = ma_y = 0 \text{ N} \qquad \sum F_x = -0.8n = ma_x$$

The y-component equation gives $n = mg = (1500 \text{ kg})(9.8 \text{ m/s}^2)$. Substituting this into the x-component equation yields

$$(1500 \text{ kg})a_x = -0.8(1500 \text{ kg})(9.8 \text{ m/s}^2) \Rightarrow a_x = (-0.8)(9.8 \text{ m/s}^2) = -7.8 \text{ m/s}^2$$

Using the constant-acceleration kinematic equation $v_f^2 = v_i^2 + 2a_x \Delta x$, we find

$$\Delta x = -\frac{v_i^2}{2a_x} = -\frac{(40 \text{ m/s})^2}{2(-7.8 \text{ m/s}^2)} = 100 \text{ m}$$

CIRCULAR MOTION, ORBITS, AND GRAVITY

6

Q6.1. Reason: Looking down from above the player runs around the bases in a counterclockwise direction, hence the angular velocity is positive.

Assess: Note that looking from below (from under the grass) the motion would be clockwise and the angular velocity would be negative. We assumed the bird's eye view because it is standard to do so, and it is difficult to view the game from below the ground. This is akin to setting up a coordinate system with the positive *x*-axis pointing left and the positive *y*-axis pointing down; the real-life physics wouldn't change any, and the calculations of measurable quantities would produce the same results, and it might even be occasionally convenient. But unless there is a clear reason to do otherwise, the usual conventions should be your first thought.

Q6.3. Reason: Acceleration is a change in *velocity*. Since velocity is a vector, it can change by changing direction, even while the magnitude (speed) remains constant. The cyclist's acceleration is *not* zero in uniform circular motion. She has a centripetal (center-seeking) acceleration.

Assess: In everyday usage, acceleration usually means only a change in speed (specifically a speeding up), hence the confusion. But in physics we must use words very carefully to communicate clearly. Everyday usage is fine outside the physics context, but while doing physics we must use the precise physics definitions of the words.

Q6.9. Reason: At the lowest point, the acceleration is upward. Thus, the tension must be greater than the weight for the net force to be upward. The tension in the string not only offsets the weight of the ball, but additionally provides the centripetal force to keep it moving in a circle.

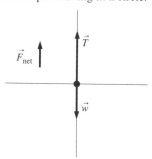

Assess: The string must have a higher strength rating than the weight of the ball in order for the ball to swing in a vertical circle.

Of course, at the top of the circle the weight itself points centripetally, so the tension in the string can be less than at the bottom.

Q6.11. Reason: **(a)** The moon's orbit around the earth is fairly circular, and it is the gravitational force of the earth on the moon that provides the centripetal force to keep the moon in its circular motion.

(b) The riders in the Gravitron carnival ride (Section 6.4) have a centripetal acceleration caused by the normal force of the walls on them.

Another example would be the biological sample in a centrifuge. The test tube walls exert a normal force on the sample toward the center of the circle.

Assess: The point is that centripetal forces are not a new *kind* of force; it is just the name we give to the force (or sum of forces) that points toward the center of the circle and keeps the object from flying off in a straight line.

Q6.13. Reason: The car is traveling along a circle and so it must have centripetal acceleration which points downward. From Newton's second law, if an object is accelerating downward, the total force on the object must be downward. The answer is C because only there is the downward force (the weight of the car) greater than the upward force (the normal force on the car) so that the total force is downward.
Assess: It makes sense since that the normal force on the car would be less than the weight of the car because, from experience, you know that you feel lighter going over a hill in your car and normal force tells you how heavy you feel. In the same way, the normal force on the car will be less than its weight.

Q6.15. Reason: The radius of the loop decreases as the carts enter and exit the loop. The centripetal acceleration is smaller for larger radius loops and larger for smaller radius loops. This means the centripetal acceleration increases from a minimum at the entry to the loop to a maximum at the top of the loop and then decreases as the cars exit the loop. This prevents a sudden change of acceleration, which can be painful. This also limits the largest accelerations to the top of the loop, so that riders only experience the maximum acceleration for a portion of the trip.
Assess: This is reasonable. If the cars entered a small radius loop directly, the centripetal acceleration would increase suddenly.

Q6.19. Reason: An object's weight is defined to be the gravitational force of the earth on the object. And the gravitational force of the earth on an object decreases with distance (as $1/r^2$), where we measure r from center to center. At the top of a mountain the climber's center is farther from the center of the earth, and so the gravitational force (i.e., the weight) is less, even though the climber's mass hasn't changed.
Assess: This is not just a change in *apparent* weight (what the scales read); this is a change in the real weight (the gravitational force).
Doubling the height of the mountain would decrease the weight by a factor of 4—but only if you take the height of the mountain to be r (from the center of the earth), *not* the height above sea level.

Problems

P6.3. Prepare: To compute the angular speed ω we use Equation 6.3 and convert to rad/s. The minute hand takes an hour to complete one revolution.
Solve:

$$\omega = \frac{\Delta\theta}{\Delta t} = \frac{1.0 \text{ rev}}{60 \text{ min}}\left(\frac{2\pi \text{ rad}}{1 \text{ rev}}\right)\left(\frac{1 \text{ min}}{60 \text{ s}}\right) = 0.0017 \text{ rad/s} = 1.7\times10^{-3} \text{ rad/s}$$

Assess: This answer applies not just to the tip, but the whole minute hand. The answer is small, but the minute hand moves quite slowly.
The second hand moves 60 times faster, or 0.10 rad/s. This too seems reasonable.

P6.5. Prepare: The airplane is to be treated as a particle in uniform circular motion on the equator around the center of the earth. We show the following pictorial representation of the problem and a list of values. To convert radians into degrees, we note that 2π rad = 360°. We will use Equation 6.4.

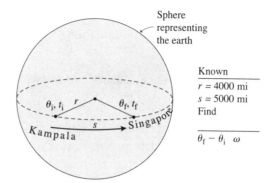

Solve: **(a)** The angle you turn through is

$$\theta_f - \theta_i = \frac{s}{r} = \frac{5000 \text{ mi}}{4000 \text{ mi}} = 1.25 \text{ rad} = 1.25 \text{ rad} \times \frac{180°}{\pi \text{ rad}} = 71.6°$$

So, the angle is 1 rad or 70°.

(b) The plane's angular speed is

$$\omega = \frac{\theta_f - \theta_i}{t_f - t_i} = \frac{1.25 \text{ rad}}{9 \text{ h}} = 0.139 \text{ rad/h} = 0.139 \frac{\text{rad}}{\text{h}} \times \frac{1 \text{ h}}{3600 \text{ s}} = 4 \times 10^{-5} \text{ rad/s}$$

Assess: An angular displacement of approximately one-fifth of a complete rotation is reasonable because the separation between Kampala and Singapore is approximately one-fifth of the earth's circumference.

P6.7. Prepare: We'll use Equation 6.4 to compute the angular displacement. We are given $\theta_i = 0.45 \text{ rad}$ and that $\Delta t = 8.0 \text{ s} - 0 \text{ s} = 8.0 \text{ s}$.

We'll do a preliminary calculation to convert $\omega = 78$ rpm into rad/s:

$$78 \text{ rpm} = 78 \frac{\text{rev}}{\text{min}} \left(\frac{2\pi \text{ rad}}{1 \text{ rev}} \right) \left(\frac{1 \text{ min}}{60 \text{ s}} \right) = 8.17 \text{ rad/s}$$

Solve: Solve Equation 6.4 for θ_f:

$$\theta_f = \theta_i + \omega \Delta t = 0.45 \text{ rad} + (8.17 \text{ rad/s})(8.0 \text{ s}) = 65.8 \text{ rad} = 10.474 \times 2\pi \text{ rad}$$
$$= 10 \times 2\pi \text{ rad} + 0.474 \times 2\pi \text{ rad} = 10 \times 2\pi \text{ rad} + 2.98 \text{ rad}$$

So the speck completed almost ten and a half revolutions. An observer would say the angular position is 3.0 rad (to two significant figures) at $t = 8.0 \text{ s}$.

Assess: Ask your grandparents if they remember the old records that turned at 78 rpm. They turned quite fast and so the music didn't last long before it was time to turn the record over.

Singles came on smaller records that turned at 45 rpm, and later "long play" (LP) records turned at 33 rpm. CDs don't have a constant angular velocity, instead they are designed to have constant linear velocity, so the motor has to change speeds. For the old vinyl records the recording had to take into account the changing linear velocity because they had constant angular velocity.

P6.11. Prepare: Use Equation 6.7 to find the speed of an object in uniform circular motion. We are given $r = 2.5 \text{ m}$ (half of the diameter).

A preliminary calculation will give ω.

$$\omega = 2\pi \text{ rad}/4.0 \text{ s} = 1.57 \text{ rad/s}$$

Solve:

$$v = \omega r = (1.57 \text{ rad/s})(2.5 \text{ m}) = 3.9 \text{ m/s}$$

Assess: A speed of 3.9 m/s seems reasonable for a merry-go-round turning this fast.

P6.15. Prepare: The pebble is a particle rotating around the axle in a circular orbit. We will use Equations 6.7 and 6.8. To convert units from rev/s to rad/s, we note that 1 rev $= 2\pi$ rad.
Solve: The pebble's angular velocity $\omega = (3.0 \text{ rev/s})(2\pi \text{ rad/rev}) = 18.85$ rad/s. The speed of the pebble as it moves around a circle of radius $r = 30$ cm $= 0.30$ m is

$$v = \omega r = (18.85 \text{ rad/s})(0.30 \text{ m}) = 5.65 \text{ m/s} = 5.7 \text{ m/s}$$

The centripetal acceleration is

$$a = \frac{v^2}{r} = \frac{(5.65 \text{ m/s})^2}{0.30 \text{ m}} = 110 \text{ m/s}^2$$

P6.19. Prepare: We are using the particle model for the car in uniform circular motion on a flat circular track. There must be friction between the tires and the road for the car to move in a circle. A pictorial representation of the car, its free-body diagram, and a list of values are shown below.

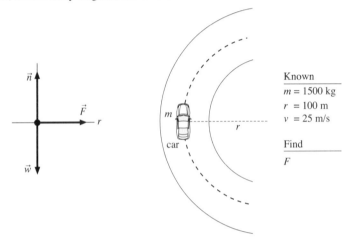

Solve: Equation 6.8 gives the centripetal acceleration

$$a = \frac{v^2}{r} = \frac{(25 \text{ m/s})^2}{100 \text{ m}} = 6.25 \text{ m/s}^2$$

The acceleration points to the center of the circle, so the net force is

$$\vec{F} = m\vec{a} = (1500 \text{ kg})(6.25 \text{ m/s}^2, \text{ toward center}) = (9400 \text{ N}, \text{ toward center})$$

This force is provided by static friction:

$$f_s = F_r = 9400 \text{ N}$$

P6.23. Prepare: Model the passenger in a roller coaster car as a particle in uniform circular motion. A pictorial representation of the car, its free-body diagram, and a list of values are shown below. Note that the normal force $\vec{n}$ of the seat pushing on the passenger is the passenger's apparent weight.

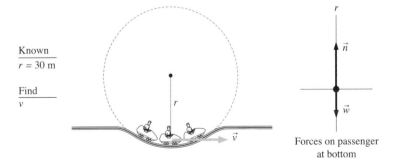

Solve: Since the passengers feel 50% heavier than their true weight, $n = 1.50\,w$. Thus, from Newton's second law, the net force at the bottom of the dip is:

$$\sum F = n - w = 1.50\,w - w = \frac{mv^2}{r} \Rightarrow 0.50\,mg = \frac{mv^2}{r} \Rightarrow v = \sqrt{0.50\,gr} = \sqrt{(0.50)(30\text{ m})(9.8\text{ m/s}^2)} = 12\text{ m/s}$$

Assess: A speed of 12 m/s or 27 mph for the roller coaster is reasonable.

P6.27. Prepare: Assume the radius of the satellite's orbit is about the same as the radius of the moon itself.

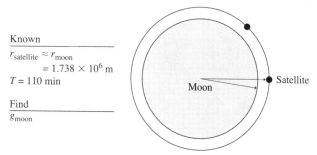

Known
_____
$r_{\text{satellite}} \approx r_{\text{moon}}$
  $= 1.738 \times 10^6$ m
$T = 110$ min
_____
Find
_____
g_{moon}

As a preliminary calculation, compute the angular velocity of the satellite:

$$w = \frac{2\pi}{T} = \frac{2\pi\text{ rad}}{110\text{ min}}\left(\frac{1\text{ min}}{60\text{ s}}\right) = 9.52 \times 10^{-4}\text{ rad/s}$$

Solve: The centripetal acceleration of the satellite is

$$a = \omega^2 r = (9.52 \times 10^{-4}\text{ rad/s})^2 (1.738 \times 10^6\text{ m}) = 1.58\text{ m/s}^2$$

Since the acceleration of a body in orbit is the local *g* experienced by that body, then this is the answer to the problem.

Assess: Our answer compares very favorably with the value of $g_{\text{Moon}} = 1.62$ m/s^2 given in the chapter.

P6.29. Prepare: Call the mass of the star *M*. Write Newton's law of gravitation for each planet.

$$F_1 = \frac{GMm_1}{r_1^2}$$

$$F_2 = \frac{GMm_2}{r_2^2} = \frac{GM(2m_1)}{(2r_1)^2}$$

Solve: Divide the two equations to get the ratio desired.

$$\frac{F_2}{F_1} = \frac{\frac{GM(2m_1)}{(2r_1)^2}}{\frac{GMm_1}{r_1^2}} = \frac{1}{2}$$

Assess: The answer is expected. Even with twice the mass, because the radius in the denominator is squared, we expect the force on planet 2 to be less than the force on planet 1.

P6.31. Prepare: Model the sun (s) and the earth (e) as spherical masses. Due to the large difference between your size and mass and that of either the sun or the earth, a human body can be treated as a particle. Use Equation 6.19.

Solve: $F_{\text{s on you}} = \dfrac{GM_s m_y}{r_{\text{s-e}}^2}$ and $F_{\text{e on you}} = \dfrac{GM_e m_y}{r_e^2}$

Dividing these two equations gives

$$\frac{F_{\text{s on y}}}{F_{\text{e on y}}} = \left(\frac{M_s}{M_e}\right)\left(\frac{r_e}{r_{\text{s-e}}}\right)^2 = \left(\frac{1.99 \times 10^{30}\text{ kg}}{5.98 \times 10^{24}\text{ kg}}\right)\left(\frac{6.37 \times 10^6\text{ m}}{1.5 \times 10^{11}\text{ m}}\right)^2 = 6.0 \times 10^{-4}$$

Assess: The result shows the smallness of the sun's gravitational force on you compared to that of the earth.

P6.35. Prepare: From Equation 6.25 we know that $T^2 \propto r^3$.

Solve: Thus, at $r_Y = 4r_X$,

$$T_Y^2 \propto (4r_X)^3 = 64r_X^3 \propto (8T_X)^2$$

With $T_Y = 8T_X$, a year on planet Y is 1600 earth days long.

Assess: This agrees perfectly with Question 6.29 where we saw that if $r_2 = 4r_1$ then $T_2 = 8T_1$. The constants in Equation 6.25 (including the mass M of the star) cancel out.

P6.39. Prepare: Model the earth (e) as a spherical mass and the satellite (s) as a point particle. The satellite has a mass is m_s and orbits the earth with a velocity v_s. The radius of the circular orbit is denoted by r_s and the mass of the earth by M_e. Use Equations 6.23 and 3.27.

Solve: The satellite experiences a gravitational force that provides the centripetal acceleration required for circular motion:

$$\frac{GM_e m_s}{r_s^2} = \frac{m_s v_s^2}{r_s} \Rightarrow r_s = \frac{GM_e}{v_s^2} = \frac{(6.67 \times 10^{-11}\,\text{N} \cdot \text{m}^2/\text{kg}^2)(5.98 \times 10^{24}\,\text{kg})}{(5500\,\text{m/s})^2} = 1.32 \times 10^7\,\text{m}$$

$$\Rightarrow T_s = \frac{2\pi R_s}{v_s} = \frac{(2\pi)(1.32 \times 10^7\,\text{m})}{(5500\,\text{m/s})} = 1.51 \times 10^4\,\text{s} = 4.2\,\text{h}$$

P6.41. Prepare: Since the speed is constant the acceleration tangent to the path at each point is zero.

Solve: Since $a = v^2/r$ and v is constant, we see that the radius of curvature of the road at point A is about three times larger than the radius of curvature at point C, so the car's centripetal acceleration at point C is three times larger than at point A.

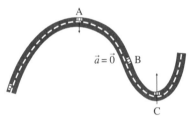

At point B there is no curvature, so there is no centripetal acceleration.

Assess: When you drive on windy roads you know that the tighter the curve the more acceleration you feel, and it is often wise to *not* keep your speed constant. Slowing down for tight curves keeps the centripetal acceleration manageable (it must be produced by the centripetal force of friction of the road on the tires).

P6.45. Prepare: Model the ball as a particle which is in a vertical circular motion. A visual overview of the ball's vertical motion is shown in the following pictorial representation, free-body diagram, and list of values. The tension in the string causes the centripetal acceleration needed for the ball's circular motion.

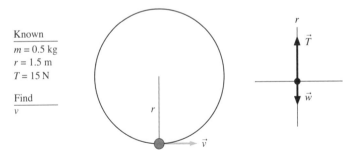

Solve: At the bottom of the circle,

$$\sum F_{\text{bottom}} = T - w = \frac{mv^2}{r} \Rightarrow (15\,\text{N}) - (0.5\,\text{kg})(9.8\,\text{m/s}^2) = \frac{(0.5\,\text{kg})v^2}{(1.5\,\text{m})} \Rightarrow v = 5.5\,\text{m/s}$$

Assess: A speed of 5.5 m/s or 12 mph is reasonable for the ball attached to a string.

P6.49. Prepare: Treat the ball as a particle in circular motion. A visual overview of the ball's circular motion is shown below in a pictorial representation, a free-body diagram, and a list of values. The mass moves in a *horizontal* circle of radius $r = 20$ cm. A component of the tension in the string toward the center of the circle causes the centripetal acceleration needed for circular motion. The acceleration $\vec{a}$ and the net force vector point to the center of the circle, *not* along the string. The other two forces are the string tension $\vec{T}$, which does point along the string, and the weight $\vec{w}$.

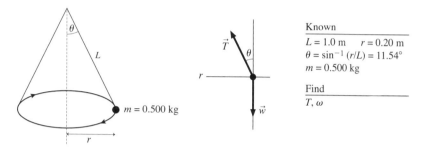

Known
$L = 1.0$ m $r = 0.20$ m
$\theta = \sin^{-1}(r/L) = 11.54°$
$m = 0.500$ kg

Find
T, ω

Solve: **(a)** Newton's second law for circular motion is

$$\sum F_y = T\cos\theta - w = T\cos\theta - mg = 0 \text{ N} \qquad \sum F_x = T\sin\theta = \frac{mv^2}{r}$$

From the *y*-equation,

$$T = \frac{mg}{\cos\theta} = \frac{(0.5 \text{ kg})(9.8 \text{ m/s}^2)}{\cos 11.54°} = 5.0 \text{ N}$$

(b) We can find the rotation speed from the *x*-equation:

$$v = \sqrt{\frac{rT\sin\theta}{m}} = 0.633 \text{ m/s}$$

The rotation frequency is $\omega = v/r = 3.165$ rad/s. Converting to rpm,

$$\omega = 3.165 \frac{\text{rad}}{\text{sec}} \times \frac{60 \text{ sec}}{1 \text{ min}} \times \frac{1 \text{ rev}}{2\pi \text{ rad}} = 30 \text{ rpm}$$

Assess: One revolution in two seconds is reasonable.

P6.51. Prepare: Since the hanging block is at rest, the total force on it is zero. The two forces are the tension in the string, T, and the weight of the puck, $-mg$. Since the revolving puck is moving at constant speed in a circle, the total force on the puck is the centripetal force. We must write the equations and solve them.
Solve: The total force on the block is $T - mg$. From Newton's second law, the total force is zero so we write:

$$T = mg = (1.20 \text{ kg})(9.8 \text{ m/s}^2) = 11.8 \text{ N}$$

The centripetal acceleration of the puck is caused by the tension in the string, so $mv^2/r = T$. We solve this to obtain:

$$v = \sqrt{Tr/m} = \sqrt{(11.8 \text{ N})(0.50 \text{ m})/(0.20 \text{ kg})} = 5.4 \text{ m/s}$$

The puck must rotate at a speed of 5.4 m/s.
Assess: It is remarkable that a block can be supported by a puck moving horizontally. But both the puck and the block are able to pull on the string—the block pulls downward on one end and the puck pulls outward on the other end. The relatively small mass of the puck is compensated by its high speed of 5.4 m/s.

P6.53. Prepare: Treat the car as a particle which is undergoing circular motion. The car is in circular motion with the center of the circle below the car. A visual overview of the car's circular motion is shown below in the following pictorial representation, free-body diagram, and list of values.

Solve: Newton's second law at the top of the hill is

$$F_{net} = \Sigma F_y = w - n = mg - n = ma = \frac{mv^2}{r} \Rightarrow v^2 = r\left(g - \frac{n}{m}\right)$$

This result shows that maximum speed is reached when $n = 0$ and the car is beginning to lose contact with the road. Then,

$$v_{max} = \sqrt{rg} = \sqrt{(50 \text{ m})(9.8 \text{ m/s}^2)} = 22 \text{ m/s}$$

Assess: A speed of 22 m/s is equivalent to 50 mph, which seems like a reasonable value.

P6.57. Prepare: We expect the centripetal acceleration to be very large because ω is large. This will produce a significant force even though the mass difference of 10 mg is so small.
A preliminary calculation will convert the mass difference to kg: 10 mg = 1.0×10^{-5} kg. If the two samples are equally balanced then the shaft doesn't feel a net force in the horizontal plane. However, the mass difference of 10 mg is what causes the force.
We'll do another preliminary calculation to convert $\omega = 70,000$ rpm into rad/s.

$$78 \text{ rpm} = 70,000\frac{\text{rev}}{\text{min}}\left(\frac{2\pi \text{ rad}}{1 \text{ rev}}\right)\left(\frac{1 \text{ min}}{60 \text{ s}}\right) = 7330 \text{ rad/s}$$

Solve: The centripetal acceleration is given by Equation 6.9 and the net force by Newton's second law.

$$F_{net} = (\Delta m)(a) = (\Delta m)(\omega^2 r) = (1.0 \times 10^{-5} \text{ kg})(7330 \text{ rad/s})^2(0.10 \text{ m}) = 54 \text{ N}$$

Assess: As we expected, the centripetal acceleration is large. The force is not huge (because of the small mass difference) but still enough to worry about. The net force scales with this mass difference, so if the mistake were bigger it could be enough to shear off the shaft.

P6.61. Prepare: Model the planet Z as a spherical mass. We will use Equation 6.22.

Solve: (a) $g_{Z \text{ surface}} = \dfrac{GM_Z}{R_Z^2} \Rightarrow 8.0 \text{ m/s}^2 = \dfrac{(6.67 \times 10^{-11} \text{ N} \cdot \text{m}^2/\text{kg}^2)M_Z}{(5.0 \times 10^6 \text{ m})^2} \Rightarrow M_Z = 3.0 \times 10^{24} \text{ kg}$

(b) Let h be the height above the north pole. Thus,

$$g_{\text{above N pole}} = \frac{GM_Z}{(R_Z + h)^2} = \frac{GM_Z}{R_Z^2\left(1 + \frac{h}{R_Z}\right)^2} = \frac{g_{Z \text{ surface}}}{\left(1 + \frac{h}{R_Z}\right)^2} = \frac{8.0 \text{ m/s}^2}{\left(1 + \frac{10.0 \times 10^6 \text{ m}}{5.0 \times 10^6 \text{ m}}\right)^2} = 0.89 \text{ m/s}^2$$

P6.63. Prepare: According to the discussion in Section 6.3, the maximum walking speed is $v_{max} = \sqrt{gr}$. The astronaut's leg is about 0.70 m long whether on earth or on Mars, but g will be difficult. Use Equation 6.22 to find g_{Mars}.
We look up the required data in the astronomical table: $m_{Mars} = 6.42 \times 10^{23}$ kg, and $R_{Mars} = 3.37 \times 10^6$ m. In part **(b)** we'll make the same assumption as in the text: The length of the leg $r = 0.70$ m.

Solve: (a)

$$g_{Mars} = \frac{GM_{Mars}}{(R_{Mars})2} = \frac{(6.67 \times 10^{-11} \text{ N} \cdot \text{m}^2/\text{kg}^2)(6.42 \times 10^{23} \text{ kg})}{(3.37 \times 106 \text{ m})^2} = 3.77 \text{ m/s}^2 \approx 3.8 \text{ m/s}^2$$

(b)

$$v_{max} = \sqrt{gr} = \sqrt{(3.77 \text{ m/s}^2)(0.70 \text{ m})} = 1.6 \text{ m/s}$$

Assess: The answer is about 3.6 mph, or about 60% of the speed the astronaut could walk on the earth. This is reasonable on a smaller celestial body. Astronauts may adopt a hopping gait like some did on the moon. Carefully analyze the units in the preliminary calculation to see that g ends up in m/s^2 or N/kg.

P6.65. Prepare: We place the origin of the coordinate system on the 20 kg sphere (m_1). The sphere (m_2) with a mass of 10 kg is 20 cm away on the x-axis, as shown below. The point at which the net gravitational force is zero must lie between the masses m_1 and m_2. This is because on such a point, the gravitational forces due to m_1 and m_2 are in opposite directions. As the gravitational force is directly proportional to the two masses and inversely proportional to the square of distance between them, the mass m must be closer to the 10-kg mass. The small mass m, if placed either to the left of m_1 or to the right of m_2, will experience gravitational forces from m_1 and m_2 pointing in the same direction, thus always leading to a nonzero force.

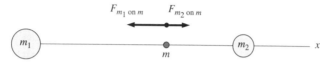

Solve:

$$F_{m_1 \text{ on } m} = F_{m_2 \text{ on } m} \Rightarrow G\frac{m_1 m}{x^2} = G\frac{m_2 m}{(0.20-x)^2} \Rightarrow \frac{20}{x^2} = \frac{10}{(0.20-x)^2} \Rightarrow 10x^2 - 8x + 0.8 = 0$$

$$\Rightarrow x = 0.683 \text{ m and } 0.117 \text{ m}$$

The value $x = 68.3$ cm is unphysical in the current situation, since this point is not between m_1 and m_2. Thus, the point $(x, y) = (11.7 \text{ cm}, 0 \text{ cm}) = (12 \text{ cm}, 0 \text{ cm})$ is where a small mass is to be placed for a zero gravitational force.

P6.71. Solve: (a) At what speed does a 1500 kg car going over a hill with a radius of 200 m have an apparent weight of 11,760 N?

(b) $2940 \text{ N} = \dfrac{1500 \text{ kg } v^2}{200 \text{ m}} \Rightarrow v = 19.8 \text{ m/s}$

ROTATIONAL MOTION

Q7.1. Reason: By convention, clockwise rotations are negative and counterclockwise rotations are positive. As a result, an angular acceleration that decreases/increases a negative angular velocity is positive/negative. In like manner, an angular acceleration that decreases/increases a positive angular velocity is negative/positive. Knowing this we can establish the situation for each figure. Figure (a) the pulley is rotating clockwise ($\omega = -$), however since the large mass is on the left it is decelerating ($\alpha = +$).

Figure (b) the pulley is rotating counterclockwise ($\omega = +$) and since the large mass is on the left it is accelerating ($\alpha = +$).

Figure (c) the pulley is rotating clockwise ($\omega = -$) and since the large mass is on the right it is accelerating ($\alpha = -$).

Figure (d) the pulley is rotating counterclockwise ($\omega = +$), however since the large mass is on the right it is decelerating ($\alpha = -$).

Assess: It is important to know the sign convention for all physical quantities that are vectors. This is especially important when working with rotational motion.

Q7.3. Reason: The question properly identified where the torques are computed about (the hinge). Torques that tend to make the door rotate counterclockwise in the diagram are positive by convention (general agreement) and torques that tend to make the door rotate clockwise are negative.

(a)   +

(b)   −

(c)   +

(d)   −

(e)   0

Assess: Looking at the diagram we see that $\vec{F}_a$ and $\vec{F}_c$ are parallel and are both creating a negative or counterclockwise torque. But since $\vec{F}_c$ is farther from the hinge, its torque will be greater. A similar argument can be made for $\vec{F}_b$ and $\vec{F}_d$. Note that $\vec{F}_e$ causes no torque since it has no moment arm.

Q7.7. Reason: The torque you exert on the branch is your weight multiplied by the lever arm, or the distance along the branch from the trunk, so the farther out you are, the greater the torque you exert on the branch.

Assess: It is a good thing that tree branches themselves get thinner away from the trunk, so the weight of the branch itself doesn't break it away from the trunk.

Q7.9. Reason: As suggested by the figure, we will assume that the larger sphere is more massive. Then the center of gravity would be at point 1, because if we suspend the dumbbell from point 1 the counterclockwise torque due to the large sphere (large weight times small lever arm) will be equal to the clockwise torque due to the small sphere (small weight times large lever arm).

Assess: Look at the figure and mentally balance the dumbbell on your finger; your finger would have to be at point 1.

The sun-earth system is similar to this except that the sun's mass is so much greater than the earth's that the center of mass (called the barycenter for astronomical objects orbiting each other) is only 450 km from the center of the sun.

Q7.13. Reason: We will assume that the table has enough friction that the ends of the rods don't slip. We will also ignore air resistance.

The angular acceleration is the quantity of interest in this question. If the two rods have the same angular acceleration, they will hit the table at the same time if released at the same time from the same angle.

However, the angular acceleration is dependent on the torque and the moment of inertia: $\alpha = \tau_{net}/I$. The net torque about the end of the rod on the table is the torque due to the weight of the rod ($w = Mg$) acting at the center of mass ($L/2$ from the end), because the normal and friction forces of the table on the end of the rod will produce a zero torque since the lever arm is zero.

Both the heavy steel rod and the pencil will be modeled as a thin rod so that the moment of inertia (given in Table 7.4) is $I = \frac{1}{3} ML^2$, but of course neither M nor L is the same in the two cases.

Now it is important to watch the mass (as well as one L) cancel out of the calculation for α.

$$\alpha = \frac{\tau_{net}}{I} = \frac{r(F)\sin\phi}{\frac{1}{3}ML^2} = \frac{r(w)\sin\phi}{\frac{1}{3}ML^2} = \frac{(\frac{L}{2})(Mg)\sin\phi}{\frac{1}{3}ML^2} = \frac{(\frac{1}{2})g\sin\phi}{\frac{1}{3}L} = \frac{3g\sin\phi}{2L}$$

What do we notice about the result?

1. M canceled, so the mass of the rod does not affect the angular acceleration. This is reminiscent of free fall where the mass also cancels in $a = F_{net}/m = mg/m$.

2. Unlike free fall, since ϕ (the angle the rod makes with the vertical) changes as the rod falls, the acceleration is not constant. In our case ϕ started out at 15° and increases to 90°; $\sin\phi$ also increases over that interval so the acceleration increases as the rod falls.

3. One L canceled, but one is still left in the denominator; the longer the rod the smaller the angular acceleration. This is the point that answers the question. Since the 1.0 m rod is longer than the 0.15 m pencil, it will have less angular acceleration and hence hit the table last.
   The pencil hits first.

Assess: The intuition we've developed about Galileo's law of falling bodies (all bodies in free fall have the same acceleration) doesn't quite apply in this case. Of course the rods aren't in free fall; it is the way the L didn't cancel in τ_{net}/I that makes the accelerations not the same.

Q7.15. Reason: The bottom of the tire, the point in contact with the road, has (relative to the ground) an instantaneous velocity of zero, but the top of the tire has an instantaneous forward velocity twice as fast as the car's forward velocity. See Figure 7.36.

When the pebble works loose it flies off tangentially forward 60 mph *faster* than the car is going. So it hits the wheel well with a relative speed of about 60 mph.

Assess: Yes, gravity pulls the pebble down and the trajectory is parabolic, but it is going so fast that it hits the wheel well before it falls far.

Problems

P7.1. Prepare: We'll assume a constant acceleration during the one revolution. We'll use the second and third equations for circular motion in Table 7.2, the third to find α and then the second to find ω_f.

Known
$r = \frac{1}{2}D = .90$ m
$\Delta t = 1.0$ s
$\Delta\theta = 1.0$ rev $= 2\pi$ rad
Find
$v_f = r\omega_f$

Solve: The third equation in Table 7.2 allows us to solve for α. That $\omega_0 = 0$ makes it easier.

$$\Delta\theta = \frac{1}{2}\alpha(\Delta t)^2$$

$$\alpha = \frac{2\Delta\theta}{(\Delta t)^2} = \frac{2(2\pi \text{ rad})}{(1.0 \text{ s})^2} = 12.6 \text{ rad/s}^2$$

The second equation in Table 7.2 gives ω_f:

$$\omega_f = \omega_0 + \alpha \Delta t = 0 \text{ rad/s} + (12.6 \text{ rad/s}^2)(1.0 \text{ s}) = 12.6 \text{ rad/s}$$

Finally, we compute $v_f = r\omega_f = (.90 \text{ m})(12.6 \text{ rad/s}) = 11 \text{ m/s}$.

Assess: This speed seems reasonable, about 1/4 of a baseball fast pitch. The hammer throw is similar to the discus, but the weight is on a wire so the radius of the circular motion is a bit longer than the arm and the release speed is a bit larger, hence the distance it goes before landing is a few meters more.

P7.5. Prepare: The magnitude of the torque in each case is $\tau = rF$ because $\sin \phi = 1$.

Solve: $\tau_1 = rF$ $\tau_2 = r2F$ $\tau_3 = 2rF$ $\tau_4 = 2r2F$

Examining the above we see that $\tau_1 < \tau_2 = \tau_3 < \tau_4$. Since for each case $\tau = rF$ (because $\sin \phi = 1$), in order to determine the torque we have just kept track of each force (F), the magnitude of the position vector (r) which locates the point of application of the force, and finally the product (rF).

Assess: As expected, both the force, and the lever arm contribute to the torque. Larger forces and larger lever arms make larger torques. Case 4 has both the largest force and the largest lever arm, hence the largest torque.

P7.7. Prepare: Torque by a force is defined as $\tau = Fr \sin \phi$ [Equation 7.5], where ϕ is measured counterclockwise from the $\vec{r}$ vector to the $\vec{F}$ vector. The radial line passing through the axis of rotation is shown below by broken line. We see that the 20 N force makes an angle of $+90°$ relative to the radius vector r_2, but the 30 N force makes an angle of $-90°$ relative to r_1.

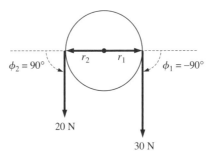

Solve: The net torque on the pulley about the axle is the torque due to the 30 N force plus the torque due to the 20 N force:

$$(30 \text{ N})r_1 \sin \phi_1 + (20 \text{ N})r_2 \sin \phi_2 = (30 \text{ N})(0.02 \text{ m})\sin(-90°) + (20 \text{ N})(0.02 \text{ m})\sin(90°)$$

$$= (-0.60 \text{ N} \cdot \text{m}) + (0.40 \text{ N} \cdot \text{m}) = -0.20 \text{ N} \cdot \text{m}$$

Assess: A negative torque will cause a clockwise rotation of the pulley.

P7.9. Prepare: The height, thickness, and mass of the door are all irrelevant (for this problem, but the mass is important for Problem 7.31). If the door closer exerts a torque of 5.2 N · m, then you need to also apply a torque of 5.2 N · m in the opposite direction. The way to do that with the least force is to make r as big as possible (the entire width of the door), and make sure the angle $\phi = 90°$.

Solve: From $\tau = rF \sin \phi$ solve for F. Then we see that the needed torque is produced with the smallest force by maximizing r and $\sin \phi$.

$$F = \frac{\tau}{r \sin \phi} = \frac{5.2 \text{ N} \cdot \text{m}}{(0.91 \text{ m})\sin 90°} = 5.7 \text{ N}$$

Assess: It is good to have problems where more than the required information is given. Part of learning to solve real-world problems is knowing (or learning) which quantities are significant, which are irrelevant, and which are negligible. Of course, in this case the mass is used later in Problem 7.31.

The answer of 5.7 N seems like a reasonable amount of force, which might be supplied, say, by a doorstop. If your doorstop is a simple wedge of wood inserted under the door (as they are at my college), you can see that it should be positioned near the outside edge of the door so the friction force will produce enough torque to keep the door open.

P7.15. Prepare: We'll set up the coordinate system so the barbell lies along the *x*-axis with the origin at the left end and use Equation 7.7 to determine the *x*-coordinate of the center of gravity. We'll model each weight (and the barbell itself in part (**b**)) as a particle, as if the mass were concentrated at each item's own center of gravity.
Solve: (**a**)

$$x_{cg} = \frac{x_1 m_1 + x_2 m_2}{m_1 + m_2} = \frac{(0.0 \text{ m})(20 \text{ kg}) + (1.7 \text{ m})(35 \text{ kg})}{20 \text{ kg} + 35 \text{ kg}} = 1.08 \text{ m} \approx 1.1 \text{ m}$$

(**b**) In this part we include the barbell (call it particle 3), whose own center of gravity is at its geometrical center ($x_3 = 1.7 \text{ m}/2 = 0.85 \text{ m}$), since we assume it has uniform density.

$$x_{cg} = \frac{x_1 m_1 + x_2 m_2 + x_3 m_3}{m_1 + m_2 + m_3} = \frac{(0.0 \text{ m})(20 \text{ kg}) + (1.7 \text{ m})(35 \text{ kg}) + (0.85 \text{ m})(8.0 \text{ kg})}{20 \text{ kg} + 35 \text{ kg} + 8.0 \text{ kg}} = 1.25 \text{ m} \approx 1.2 \text{ m}$$

Assess: To two significant figures the answers to both parts are close. Taking the barbell into account didn't move the center of gravity much for two reasons: It wasn't very massive, and its center of mass was already near the center of mass of the system.

P7.17. Prepare: How will you estimate the mass of your arm? Of course different people's arms have different masses, but even different methods of estimating the mass of your specific arm will produce slightly different results. But it is a good exercise, even if we have only one significant figure of precision. One way would be to make some rough measurements. Because the density of your arm is about the density of water, you could fill a garbage can to the brim with water, insert your arm, and weigh the water that overflows. Or you could look at Figure 7.24, which indicates that a guess of $m = 4.0 \text{ kg}$ for a person whose total mass is 80 kg is good.
The gravitational force on a 4.0 kg object is $w = \text{mg} (4.0 \text{ kg})(10 \text{ m/s}^2) = 40 \text{ N}$.
I know it is about 1 yd from the tip of my outstretched arm to my forward-facing nose, but a meter is a bit bigger than a yard, so I would estimate that from fingertip to shoulder would be about 0.7 m. If I model my arm as a uniform cylinder of uniform density then the center of gravity would be at its center—0.35 m from the shoulder joint (the "hinge"). If I further refine the model of my arm, it is heavier nearer the shoulder and lighter toward the hand, so I will round the location of its center of gravity down to 0.30 m from the shoulder.
Solve: Since the arm is held horizontally $\sin\phi = 1$.

$$\tau = rF \sin\phi = (0.30 \text{ m})(40 \text{ N})(1) = 12 \text{ N} \cdot \text{m}$$

Assess: Your assumptions and estimates (and, indeed, your arm) might be different, but your answer will probably be in the same order of magnitude; you probably won't end up with an answer 10 × bigger or 10 × smaller.

P7.19. Prepare: First let's divide the object into two parts. Let's call part #1 the part to the left of the point of interest and part #2 the part to the right of the point of interest. Next using our sense of center of gravity, we know the cm of part #1 is at 12.5 m and the cm of part #2 is at +37.5 cm. We also know the mass of part #1 is one fourth the total mass of the object and the mass of part #2 is three fourths the total mass of the object. Finally, we can determine the gravitational torque of each part using any of the three expressions for torque as shown below:
Equation (7.3) $\tau = rF_\perp$ is straightforward to use because the forces are perpendicular to the position vectors, which locate the point of application of the force.
Equation (7.4) $\tau = r_\perp F$ is straightforward to use because the position vectors that locate the point of application of the forces are also the moment arms for the forces.
Equation (7.4) $\tau = rF \sin\phi$ is straightforward to use because the angles are either $90°$ or $270°$
Solve: Using equation (7.4) we obtain the following

$$\tau_{net} = \tau_1 + \tau_2 = m_1 g r_1 \sin(90°) + m_2 g r_2 \sin(-90°)$$

$$= (0.5 \text{ kg})(9.8 \text{ m/s}^2)(-0.125 \text{ m})(1) + (0.75 \text{ kg})(9.8 \text{ m/s}^2)(0.375 \text{ m})(-1) = -2.1 \text{ N} \cdot \text{m}$$

Assess: According to this answer, if released the object should rotate in a clockwise direction. Looking at the figure this is exactly what we would expect to happen.

P7.25. Prepare: A table tennis ball is a spherical shell, and we look that up in Table 7.4. The radius is half the diameter.
Solve:

$$I = \frac{2}{3}MR^2 = \frac{2}{3}(0.0027\,\text{kg})(0.020\,\text{m})^2 = 7.2 \times 10^{-7}\,\text{kg} \cdot \text{m}^2$$

Assess: The answer is small, but then again, it isn't hard to start a table tennis ball rotating or stop it from doing so. By the way, this calculation can be done in one's head without a calculator by writing the data in scientific notation and mentally keeping track of the significant figures:

$$I = \frac{2}{3}(2.7 \times 10^{-3}\,\text{kg})(2.0 \times 10^{-2}\,\text{m})^2 = \frac{2}{3}(27 \times 10^{-4}\,\text{kg})(2 \times 10^{-2}\,\text{m})^2 = \frac{27}{3} \cdot 2 \cdot 2^2 \times 10^{-4} \times 10^{-4}\,\text{kg} \cdot \text{m}^2$$
$$= 9 \cdot 8 \times 10^{-8}\,\text{kg} \cdot \text{m}^2 = 72 \times 10^{-8}\,\text{kg} \cdot \text{m}^2 = 7.2 \times 10^{-7}\,\text{kg} \cdot \text{m}^2$$

P7.27. Prepare: When the problem says the connecting rods are "very light" we know to ignore them (especially since we aren't given their mass).
Since all of the mass (seats plus children) is equally far from the axis of rotation we simply use $I = MR^2$ where M is the total mass.
Solve:

$$I = MR^2 = [4(5.0\,\text{kg}) + 15\,\text{kg} + 20\,\text{kg}](1.5\,\text{m})^2 = 120\,\text{kg} \cdot \text{m}^2$$

Assess: When all of the mass is the same distance R from the axis of rotation, the calculation of I is relatively simple.

P7.29. Prepare: Treat the bicycle rim as a hoop, and use the expression given in Table 7.4 for the moment of inertia of a hoop. Manipulate this expression to obtain the mass.
Solve: The mass of the rim is determined by

$$m = I / R^2 = 0.19\,\text{kg} \cdot \text{m}^2 / (0.65\,\text{m} / 2)^2 = 1.8\,\text{kg}$$

Assess: Note that the units reduce to kg as expected. This amount seems a little heavy, but since we are not told what type of bicycle it is not an unreasonable amount.

P7.31. Prepare: We will assume we are to compute I about an edge (the hinge of the door), so from Table 7.4 we have $I = \frac{1}{3}Ma^2$ where $a = 0.91\,\text{m}$ and $M = 25\,\text{kg}$.
Solve: (a)

$$I = \frac{1}{3}Ma^2 = \frac{1}{3}(25\,\text{kg})(0.91\,\text{m})^2 = 6.9\,\text{kg} \cdot \text{m}^2$$

(b) After we let go of the door (or remove the doorstop) the net torque is the torque due to the door closer, which Problem 7.9 says is $5.2\,\text{N} \cdot \text{m}$.

$$\alpha = \frac{\tau_{\text{net}}}{I} = \frac{5.2\,\text{N} \cdot \text{m}}{6.9\,\text{kg} \cdot \text{m}^2} = 0.75\,\text{rad/s}^2$$

Assess: The results seem to be in a reasonable range. As long as the net torque of $0.52\,\text{N} \cdot \text{m}$ continues to act, the door will continue to accelerate. Usually there are dampeners to exert a slowing torque so that the door doesn't slam.

P7.35. Prepare: A circular plastic disk rotating on an axle through its center is a rigid body. Assume the axis is perpendicular to the disk. Since $\tau = I\alpha$ is the rotational analog of Newton's second law $F = ma$, we can use this relation to find the net torque on the object. To determine the torque (τ) needed to take the plastic disk from $\omega_i = 0\,\text{rad/s}$ to $\omega_f = 1800\,\text{rpm} = (1800)(2\pi)/60\,\text{rad/s} = 60\pi\,\text{rad/s}$ in $t_f - t_i = 4.0\,\text{s}$, we need to determine the angular acceleration (α) and the disk's moment of inertia (I) about the axle in its center. The radius of the disk is $R = 10.0\,\text{cm}$.

Solve: We have

$$I = \frac{1}{2} MR^2 = \frac{1}{2}(0.200 \text{ kg})(0.10 \text{ m})^2 = 1.0 \times 10^{-3} \text{ kg} \cdot \text{m}^2$$

$$\omega_f = \omega_i + \alpha(t_f - t_i) \Rightarrow \alpha = \frac{\omega_f - \omega_i}{t_f - t_i} = \frac{60\pi \text{ rad/s} - 0 \text{ rad/s}}{4.0 \text{ s}} = 15\pi \text{ rad/s}^2$$

Thus, $\tau = I\alpha = (1.0 \times 10^{-3} \text{ kg} \cdot \text{m}^2)(15\pi \text{ rad/s}^2) = 0.047 \text{ N} \cdot \text{m}$.

Assess: The solution to this problem required a knowledge of torque, moment of inertia, rotational dynamics and rotational kinematics. You should consider it an accomplishment to have mastered these concepts and then combined them to solve a problem.

P7.37. Prepare: What causes angular accelerations? (Net) torques. We'll apply the rotational version of Newton's second law. We'll write the net torque as $\Sigma\tau$ to emphasize that we are summing the two given torques; the 12 N force is producing a positive (counterclockwise) torque, while the 10 N force is producing a negative (clockwise) torque. For each torque $R = 0.30$ m.

We will assume that the rope comes off tangent to the pulley on each side, so that $\phi = 90°$ and $\sin\phi = 1$.

Looking up the formula for I of a cylinder in Table 7.4, and using $M = 0.80$ kg, gives

$$I = \frac{1}{2} MR^2 = \frac{1}{2}(0.80 \text{ kg})(0.30 \text{ m})^2 = 0.036 \text{ kg} \cdot \text{m}^2$$

Solve:

$$\alpha = \frac{\Sigma\tau}{I} = \frac{(0.30 \text{ m})(12 \text{ N}) - (0.30 \text{ m})(10 \text{ N})}{0.036 \text{ kg} \cdot \text{m}^2} = \frac{(0.30 \text{ m})(12 \text{ N} - 10 \text{ N})}{0.036 \text{ kg} \cdot \text{m}^2} = 17 \text{ rad/s}^2$$

Assess: This result answers the question. The proper units cancel to give α in rad/s².

Notice that the specific angles the ropes make with the vertical do not matter, as long as they are exerting torques in opposite directions and coming off of the pulley tangentially.

P7.43. Prepare: The component of the static friction force that keeps the pebble in circular motion is centripetal and the centripetal force is determined by $F_c = mv^2 / R$.

We must remember that the top of the tire is going twice as fast as the car (see Figure 7.36); therefore the pebble will be released when it is at the top.

We are given $m = 0.0012$ kg, $r = D/2 = 0.36$ m, and the central frictional force $F = 3.6$ N.

Solve: At the point just before release $F = ma = m\frac{v^2}{r}$. Solve this for v (the speed of the top of the tire).

$$v = \sqrt{Fr/m} = \sqrt{(3.6 \text{ N})(0.38 \text{ m})/0.0012 \text{ kg}} = 33.8 \text{ m/s}$$

The car is going half the speed of the pebble, or 17 m/s.
Assess: This is equal to 38 mph, which is a reasonable speed.

P7.45. Prepare: The crankshaft is a rotating rigid body. The crankshaft's angular acceleration is given as $\alpha = \Delta\omega/\Delta t$, or slope of the angular-velocity-versus-time graph.

Solve: The crankshaft at $t = 0$ s has an angular velocity of 250 rad/s. It gradually slows down to 50 rad/s in 2 s, maintains a constant angular velocity for 2 s until $t = 4$ s, and then speeds up to 200 rad/s from $t = 4$ s to $t = 7$ s. The angular acceleration (α) graph is based on the fact that α is the slope of the ω-versus-t graph.

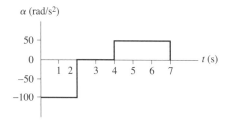

Assess: Knowing that the slope of the angular velocity-versus-time plot is the angular acceleration, we can establish a plot of angular acceleration-versus-time.

P7.49. Prepare: The disk is a rotating rigid body and it rotates on an axle through its center. We will use Equation 7.5 to find the net torque.

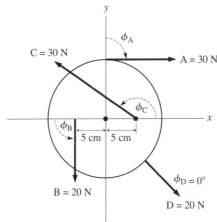

Solve: The net torque on the axle is

$$\tau = F_A r_A \sin \phi_A + F_B r_B \sin \phi_B + F_C r_C \sin \phi_C + F_D r_D \sin \phi_D$$

$$= (30 \text{ N})(0.10 \text{ m}) \sin (-90°) + (20 \text{ N})(0.05 \text{ m}) \sin 90° + (30 \text{ N})(0.05 \text{ m}) \sin 135° + (20 \text{ N})(0.10 \text{ m}) \sin 0°$$

$$= -3 \text{ N} \cdot \text{m} + 1 \text{ N} \cdot \text{m} + 1.0607 \text{ N} \cdot \text{m} = -0.94 \text{ N} \cdot \text{m}$$

Assess: A negative net torque means a clockwise acceleration of the disk.

P7.51. Prepare: Equation 7.9 tells us the center of gravity of a compound object. If we take all the rest of the body other than the arms as one object (call it the trunk, even though it includes head and legs) then we can write

$$y_{cg} = \frac{y_{trunk} m_{trunk} + 2 y_{arm} m_{arm}}{M}$$

where $M = m_{trunk} + m_{arm} = 70 \text{ kg}$ (the mass of the whole body).

The language "by how much does he raise his center of gravity" makes us think of writing Δy_{cg}.

Since we have modeled the arm as a uniform cylinder 0.75 m long, its own center of gravity is at its geometric center, 0.375 m from the pivot point at the shoulder. So raising the arm from hanging down to straight up would change the height of the center of gravity of the arm by twice the distance from the pivot to the center of gravity: $(\Delta y_{cg})_{arm} = 2(0.375 \text{ m}) = 0.75 \text{ m}$.

Solve:

$$\Delta(y_{cg})_{body} = (y_{cg})_{with \ arms \ up} - (y_{cg})_{with \ arms \ down} = \frac{(y_{cg})_{trunk} m_{trunk} + 2(y_{cg})_{arm, \ up} m_{arm}}{M} - \frac{(y_{cg})_{trunk} m_{trunk} + 2(y_{cg})_{arm, \ down} m_{arm}}{M}$$

$$= \frac{2 m_{arm}}{M}((y_{cg})_{arm, \ up} - (y_{cg})_{arm, \ down}) = \frac{2 m_{arm}}{M}(\Delta y_{cg})_{arm} = \frac{2(3.5 \text{ kg})}{70 \text{ kg}}(0.75 \text{ m}) = 0.075 \text{ m} = 7.5 \text{ cm}$$

Assess: 7.5 cm seems like a reasonable amount, not a lot, but not too little. The trunk term subtracted out, which is both expected and good because we didn't know $(y_{cg})_{trunk}$.

P7.55. Prepare: Since the rods are "light" we will neglect their contribution. We will also have to neglect the contribution of the ball through which the axis goes because we do not know the radius of the balls. It is likely that the other two balls contribute vastly more to the total I than the one pierced by the axis does, unless the balls are large compared to the lengths of the rods.

The other two balls will each contribute the same amount, and we will treat them as point particles (both because this is generally safe, and because we do not know their radius). See Equation 7.14. The distance r for each of them from the axis is 0.30 m; we also know the mass of each ball is $m = 0.10$ kg.

Solve:

$$I = (2)mr^2 = (2)(0.10 \text{ kg})(0.30 \text{ m})^2 = 0.018 \text{ kg} \cdot \text{m}^2$$

Assess: The balls are not very massive, nor are the rods particularly long, so we are satisfied with an answer that appears smallish. If the balls were five times as massive and they were 1 m away from the axis, then the answer would be 1 kg · m².

P7.59.   Prepare: This problem requires a knowledge of translational ($F_{\text{net}} = ma$) and rotational ($\tau_{\text{net}} = I\alpha$) dynamics. Notice that the counter clockwise torque is greater than the clockwise torque, hence the system will rotate counterclockwise. Let's agree to call any force that tends to accelerate the system positive and any force that tends to decelerate the system negative. Also let's agree to call the small disk M_1 and the large disk M_2.

Solve:

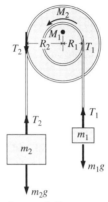

Write Newton's second law equation for m_2 and m_1 as follows:

$$m_2 g - T_2 = m_2 a_2 = m_2 R_2 \alpha \quad \text{or} \quad T_2 = m_2 g - m_2 R_2 \alpha$$

and

$$T_1 - m_1 g = m_1 a_1 = m_1 R_1 \alpha \quad \text{or} \quad T_1 = m_1 g + m_1 R_1 \alpha$$

The net torque acting on the system may be determined by

$$\tau = R_2 T_2 - R_1 T_1 = R_2 (m_2 g - m_2 \alpha R_2) - R_1 (m_1 g + m_1 \alpha R_1)$$

The moment of inertia of the system is

$$I = I_1 + I_2 = (M_1 R_1^2 / 2) + (M_2 R_2^2 / 2)$$

Knowing

$$\tau = I\alpha$$

We may combine the above to get

$$R_2 (m_2 g - m_2 \alpha R_2) - R_1 (m_1 g + m_1 \alpha R_1) = [(M_1 R_1^2 / 2) + (M_2 R_2^2 / 2)]\alpha$$

Which may be solved for α to obtain

$$\alpha = \frac{(m_2 R_2 - m_1 R_1)g}{R_2^2 (m_2 + M_2 / 2) + R_1^2 (m_1 + M_1 / 2)} = 3.5 \text{ rad/s}^2$$

Assess: This angular acceleration amounts to speeding up about a half revolution per second every second. That is not an unreasonable amount.

P7.61. Prepare: Two balls connected by a rigid, massless rod are a rigid body rotating about an axis through the center of gravity. Assume that the size of the balls is small compared to 1 m. We have placed the origin of the coordinate system on the 1.0 kg ball. Since $\tau = I_{\text{about cg}}\alpha$, we need the moment of inertia and the angular acceleration to be able to calculate the required torque.

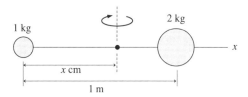

Solve: The center of gravity and the moment of inertia are

$$x_{\text{cm}} = \frac{(1.0 \text{ kg})(0 \text{ m}) + (2.0 \text{ kg})(1.0 \text{ m})}{(1.0 \text{ kg} + 2.0 \text{ kg})} = 0.667 \text{ m} \quad \text{and} \quad y_{\text{cm}} = 0 \text{ m}$$
$$I_{\text{about cm}} = \sum m_i r_i^2 = (1.0 \text{ kg})(0.667 \text{ m})^2 + (2.0 \text{ kg})(0.333 \text{ m})^2 = 0.667 \text{ kg} \cdot \text{m}^2$$

We have $\omega_f = 0$ rad/s, $t_f - t_i = 5.0$ s, and $\omega_i = 20$ rpm $= 20(2\pi \text{ rad/60 s}) = \frac{2}{3}\pi$ rad/s, so $\omega_f = \omega_i + \alpha(t_f - t_i)$ becomes

$$0 \text{ rad/s} = \left(\frac{2\pi}{3}\text{rad/s}\right) + \alpha(5.0 \text{ s}) \Rightarrow \alpha = -\frac{2\pi}{15}\text{rad/s}^2$$

Having found I and α, we can now find the torque τ that will bring the balls to a halt in 5.0 s:

$$\tau = I_{\text{about cm}}\alpha = \left(\frac{2}{3}\text{kg} \cdot \text{m}^2\right)\left(-\frac{2\pi}{15}\text{ rad/s}^2\right) = -\frac{4\pi}{45} \text{ N} \cdot \text{m} = -0.28 \text{ N} \cdot \text{m}$$

The magnitude of the torque is 0.28 N·m.
Assess: The minus sign with the torque indicates that the torque acts clockwise.

P7.63. Prepare: The pulley is a rigid rotating body. We also assume that the pulley has the mass distribution of a disk and that the string does not slip. Because the pulley is not massless and frictionless, tension in the rope on both sides of the pulley is *not* the same. We will have to be careful with the appropriate masses when we write below Newton's second law for the blocks and the pulley. A pictorial diagram of the problem and free-body diagrams for the two blocks are shown. We have placed the origin of the coordinate system on the ground.

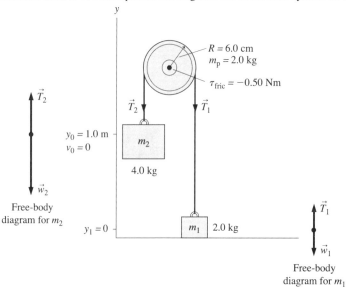

Solve: Applying Newton's second law to m_1, m_2, and the pulley yields the three equations:

$$T_1 - w_1 = m_1 a_1 \qquad -w_2 + T_2 = m_2 a_2 \qquad T_2 R - T_1 R - 0.50 \text{ N} \cdot \text{m} = I\alpha$$

Noting that $-a_2 = a_1 = a, I = \frac{1}{2} m_p R^2$, and $\alpha = a/R$, the above equations simplify to

$$T_1 - m_1 g = m_1 a \qquad m_2 g - T_2 = m_2 a \qquad T_2 - T_1 = \left(\frac{1}{2} m_p R^2\right)\left(\frac{a}{R}\right)\frac{1}{R} + \frac{0.50 \text{ N} \cdot \text{m}}{R} = \frac{1}{2} m_p a + \frac{0.50 \text{ N} \cdot \text{m}}{0.060 \text{ m}}$$

Adding these three equations,

$$(m_2 - m_1)g = a\left(m_1 + m_2 + \frac{1}{2} m_p\right) + 8.333 \text{ N} \Rightarrow a = \frac{(m_2 - m_1)g - 8.333 \text{ N}}{m_1 + m_2 + \frac{1}{2} m_p}$$

$$= \frac{(4.0 \text{ kg} - 2.0 \text{ kg})(9.8 \text{ m/s}^2) - 8.333 \text{ N}}{2.0 \text{ kg} + 4.0 \text{ kg} + (2.0 \text{ kg}/2)} = 1.610 \text{ m/s}^2$$

We can now use kinematics to find the time taken by the 4.0 kg block to reach the floor:

$$y_f = y_i + v_i(t_f - t_i) + \frac{1}{2} a_2(t_f - t_i)^2 \Rightarrow 0 = 1.0 \text{ m} + 0 + \frac{1}{2}(-1.610 \text{ m/s}^2)(t_f - 0 \text{ s})^2 \Rightarrow t_f = \sqrt{\frac{2(1.0 \text{ m})}{(1.610 \text{ m/s}^2)}} = 1.1 \text{ s}$$

Assess: Compared to free fall where we would use $a = -9.80$ m/s^2, $a = -1.61$ m/s^2, and a time of 1.1 s for the block to reach floor are reasonable.

EQUILIBRIUM AND ELASTICITY

Q8.1. Reason: Because the definition of equilibrium includes $\Sigma\tau = 0$ as well as $\Sigma\vec{F} = \vec{0}$, an object that experiences exactly two forces that are equal in magnitude and opposite in direction may still not be in equilibrium because the forces could cause a net torque even though the sum of the forces is zero. See Figure 8.1.

Furthermore, an object could even have $\Sigma\tau = 0$ as well as $\Sigma\vec{F} = \vec{0}$ (and therefore be in equilibrium) but still not be in *static* equilibrium if it is moving at a constant velocity.

Assess: Think carefully about the definition of equilibrium, especially what it *doesn't* say.

Q8.3. Reason: The ladder could *not* be in static equilibrium. Consider the forces in the horizontal direction. There is a normal force exerted by the wall on the top of the ladder, but no other object (in the absence of friction) exerts a counterbalancing force on the ladder in the opposite direction. Examine Figure 8.9.

Assess: This makes sense from a common sense standpoint. If a ladder is about to slip out one tries to increase the friction at the point of contact with the floor, or to produce a horizontal component of a normal force by wedging it.

Q8.5. Reason: For divers to be stable on the board before the dive their center of gravity must be over an area of support, that is, over the board. Extending their arms moves the center of gravity over the board.

Assess: If the arms are not extended, then the center of gravity would be over the edge of the diving board when they stand on their toes with heels extended out. They would not be in static equilibrium and would topple over before getting off a good clean dive. The other option to get the center of gravity over the board (besides extending arms) is to lean forward slightly toward the board.

Q8.9. Reason: Before Carlos came along the wall also pulled on the spring with a 200 N force when Bob did, that is, there was a 200 N tension force all along the spring. When Carlos arrives he takes the place of the wall but the spring must still stretch 20 cm. The only difference is that now Carlos also moves whereas the wall didn't.
(a) 10 cm. Though the spring stretched 20 cm originally, its center moved by 10 cm and so Bob's end moved 10 cm away from (farther than) the center. In the tug-of-war the center stays still so Bob's end only moves 10 cm.
(b) 10 cm in the other direction. The total stretch under a 200 N tension must still be 20 cm.

Assess: These answers fit well with Hooke's law. In either case the 200 N tension produced a total stretch of 20 cm.

Q8.13. Reason: The force needed to bend a "beam," whether it's a nail or a steel wool fiber, depends on the thickness-to-length ratio. The diameter (thickness) of a steel wool fiber is *much* less, relative to its length, than that of a steel nail. Thus it takes only a very small force to bend and flex the thin fibers of steel wool, but a very large force to bend a steel nail.

Assess: Fiberglass is also flexible while a thicker glass rod is not, for the same reasons.
The extreme case is carbon nanotubes that are *so* thin that they bend easily, but if made into a solid bulk substance as thick as nails would be more resistant to bending.

Problems

P8.1. Prepare: Because the board is "very light" we will assume that it is massless and does not contribute to the scale reading, nor does it contribute any torques. The sum of the two scale readings must equal the woman's weight: $w = mg = (64 \text{ kg})(9.8 \text{ m/s}^2) = 627 \text{ N} \approx 630 \text{ N}$.

Solve: Compute the torques around the point the board rests on the left scale. The woman's weight creates a clockwise (negative) torque; and the normal force n_{right} of the right scale creates a counterclockwise (positive) torque.

$$\Sigma\tau = (2.0\ \text{m})(n_{\text{right}}) - (1.5\ \text{m})(627\ \text{N}) = 0\ \text{N}\cdot\text{m}$$

The right scale reads n_{right}:

$$n_{\text{right}} = \frac{(1.5\ \text{m})(627\ \text{N})}{2.0\ \text{m}} = 470\ \text{N}$$

By simple subtraction the left scale reads

$$n_{\text{left}} = 627\ \text{N} - 470\ \text{N} = 160\ \text{N}$$

Assess: The answer is reasonable. Since the woman is three times farther from the left scale than the right one, it (the left one) reads three times less. And the two scale readings sum to the woman's weight, as required. Not only *could* we have chosen the pivot point at the right scale and produced the same answer, but we *should* do so as a check.

$$\Sigma\tau = (0.5\ \text{m})(627\ \text{N}) - (2.0\ \text{m})(n_{\text{left}}) = 0\ \text{N}\cdot\text{m}$$

$$n_{\text{left}} = \frac{(0.5\ \text{m})(627\ \text{N})}{2.0\ \text{m}} = 160\ \text{N}$$

And so

$$n_{\text{right}} = 627\ \text{N} - 157\ \text{N} = 470\ \text{N}$$

It is true that $\Sigma\tau = 0$ around *any* point (for equilibrium), but we picked the two we did (the second as a check) because then the resulting torque equations each had only one unknown in them.

P8.3. Prepare: Compute the torques around the bottom of the right leg of the table. The horizontal distance from there to the center of gravity of the table is $\dfrac{2.10\ \text{m}}{2} = 0.55\ \text{m} = 0.50\ \text{m}$.

Solve: Call the horizontal distance from the bottom of the right leg to the center of gravity of the man x.

$$\Sigma\tau = (56\ \text{kg})(9.8\ \text{m/s}^2)(0.50\ \text{m}) - (70\ \text{kg})(9.8\ \text{m/s}^2)x = 0 \quad\Rightarrow\quad x = 0.40\ \text{m}$$

The distance from the right edge of the table is now $0.55\ \text{m} - 0.40\ \text{m} = 0.15\ \text{m} = 15\ \text{cm}$.
Assess: It seems likely that the table would tip if the man were closer than 15 cm to the edge.

P8.7. Prepare: The massless rod is a rigid body. To be in equilibrium, the object must be in both translational equilibrium ($\vec{F}_{\text{net}} = 0\ \text{N}$) and rotational equilibrium ($\tau_{\text{net}} = 0$). We have $(F_{\text{net}})_y = (40\ \text{N}) - (100\ \text{N}) + (60\ \text{N}) = 0\ \text{N}$, so the object is in translational equilibrium. Our task now is to calculate the net torque.

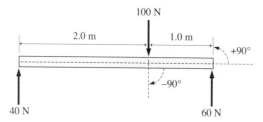

Solve: Measuring τ_{net} about the left end,

$$\tau_{\text{net}} = (60\ \text{N})(3.0\ \text{m})\sin(+90°) + (100\ \text{N})(2.0\ \text{m})\sin(-90°) = -20\ \text{N}\cdot\text{m}$$

The object is not in equilibrium.

P8.11. Prepare: The pole is in equilibrium, which means $\Sigma\tau = 0$; it is convenient to compute the torques around the bottom of the pole (as suggested by the hinge there). It is definitely worth noting that the triangles made by the wires, the ground, and the pole are congruent (rotate one triangle 90° around corner at the bottom of the pole to see this).

Solve: For $\Sigma\tau = 0$ we want the magnitude of the clockwise torque to equal the magnitude of the counterclockwise torque. The magnitude of each torque is the force (the tension in the guy wire) multiplied by the lever arm (the perpendicular distance from the line in the direction of the force to the pivot, see Equation 7.4). Examination of the figure, and realization that the triangles are congruent, leads to the realization that the lever arms are equal for the two wires.

Since the magnitudes of the torques must be equal, and the lever arms are equal, then the magnitudes of the forces must be equal. Hence the ratio of the tension in the left wire to the tension in the right wire is 1 ($T_2/T_1 = 1$).

Assess: Theoretically, the pole could be at rest in equilibrium if there were no wires, but any slight perturbation would make it fall over (an example of unstable equilibrium), and so guy wires are used which can keep the pole in equilibrium even if perturbed.

Whether it is preferable to use guy wires like the one on the left or like the one on the right may depend on how much room around the base of the pole is available, and where one would rather attach the wires to the pole.

P8.13. Prepare: The center of gravity of the magazine rack must be over the base of support to be stable. In this case the rule of thumb given in the text that "a wider base of support and/or a lower center of gravity improve stability" indicates that we expect the tipping angle to be small. The center of gravity need only move 2.5 cm horizontally for the rack to be on the verge of tipping.

On the diagram construct a right triangle by first dropping a vertical from the center of gravity to the middle of the base and drawing the hypotenuse from the center of gravity to the edge of the base. Now we have a right triangle with legs of 16 cm and 2.5 cm. The angle we desire is the small angle at the top of the triangle.

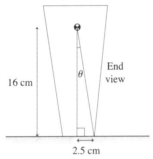

If this is not yet clear, draw a second diagram with the rack tipped just so the center of gravity is over the edge of the base of support and see that the tipping angle is the same θ that we labeled in the triangle.

Solve: The angle will be given by the arctangent of the opposite side over the adjacent side.

$$\theta = \tan^{-1}\left(\frac{2.5\,\text{cm}}{16\,\text{cm}}\right) = 8.9°$$

Assess: The angle of 8.9° is fairly small, as expected.

The precise shape of the cross section of the rack is unimportant as long as the base of support and the center of gravity are where they are.

P8.17. Prepare: Hooke's law is given in Equation 8.3, $(F_{sp})_x = -k\Delta x$. It relates the force on a spring to the stretch; the constant of proportionality is k, the spring constant that we are asked to find.

Solve: The minus sign in Equation 8.3 simply indicates that the force and the stretch are in opposite directions (that the force is a *restoring* force); k is always positive, so we'll drop the minus sign and just use magnitudes of $(F_{sp})_x$ and Δx since we would otherwise have to set up a more explicit coordinate system. See Equation 8.1.

$$k = \frac{(F_{sp})_x}{\Delta x} = \frac{25\,\text{N}}{0.030\,\text{m}} = 830\,\text{N/m}$$

Assess: This result indicates a fairly stiff spring, but certainly within reason.

P8.19. Prepare: A visual overview below shows the details, including a free-body diagram, of the problem. We will assume an ideal spring that obeys Hooke's law.

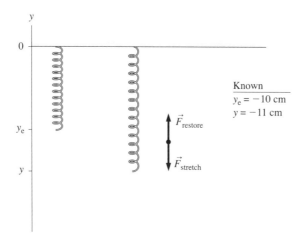

Solve: **(a)** The spring force or the restoring force is $F_{sp} = -k\Delta y$. For $\Delta y = -1.0$ cm and the force in Newtons,

$$F_{sp} = F = -k\Delta y \Rightarrow k = -F/\Delta y = -F/(-0.01\,\text{m}) = 100F\ \text{N/m}$$

Notice that Δy is negative, so F_{sp} is positive.

We can now calculate the new length for a restoring force of 3F:

$$F_{sp} = 3F = -k\Delta y = (-100\,F)\Delta y \Rightarrow \Delta y = -0.03\,\text{m}$$

From $\Delta y = y - y_e = -0.03$ m, or $y = -0.03$ m $+ y_e$, or $y = -0.03$ m $+ (-0.10$ m$) = -0.13$ m, the length of the spring is 0.13 m.

(b) The new compressed length for a restoring force of 2F can be calculated as:

$$F_{sp} = 2F = -k\Delta y = (-100\,F)\Delta y \Rightarrow \Delta y = -0.02\,\text{m}$$

Using $\Delta y = y_e - y = -0.02$ m, or $y = 0.02$ m $+ y_e$, or $y = 0.02$ m $+ (-0.10$ m$) = -0.08$ m, the length of the compressed spring is 0.08 m.

Assess: The stretch Δx is proportional to the applied force, as both parts of this problem demonstrate. Of course, this bet is off if the spring is stretched or compressed far enough to take it out of the linear region.

P8.23. Prepare: Assume that the spring is ideal and obeys Hooke's law. Also we will model the sled as a particle. The only horizontal force acting on the sled is $\vec{F}_{sp}$. A pictorial representation and a free-body diagram are shown below.

Solve: Applying Newton's second law to the sled gives

$$\Sigma(F_{\text{on sled}})_x = F_{\text{sp}} = ma_x \Rightarrow k\Delta x = ma_x \Rightarrow a_x = k\Delta x/m = (150 \text{ N/m})(0.20 \text{ m})/20 \text{ kg} = 1.5 \text{ m/s}^2$$

P8.25. Prepare: Rearrange Equation 8.6 to see that the stretch is proportional to the length (for part (**a**)) and inversely proportional to the area (for part (**b**)).

$$\Delta L = \frac{LF}{AY}$$

Solve: (**a**) In this part everything on the right side of the equation stays constant except the length L. Since the length of the second wire is twice the length of the first wire, then the second wire will stretch twice as much by the same force. So the answer is 2 mm.
(**b**) In this part everything on the right side of the equation stays constant except the cross-sectional area A. The cross-sectional area of the third wire is four times the area of the first wire, since $A = \pi r^2 = \pi(D/2)^2$ and the diameter of the third wire is twice the diameter of the first wire, so the third wire will stretch one-quarter as much by the same force. The answer is 0.25 mm.
Assess: This problem is worth mentally reviewing to make sure the explanation given makes sense, and to tuck the results away as tidbits of practical knowledge.

P8.27. Prepare: Equation 8.6 relates the quantities in question.
Look up Young's modulus for steel in Table 8.1: $Y_{\text{steel}} = 20 \times 10^{10} \text{ N/m}^2$.
Convert all length data to meters: $D = 1.0 \text{ cm} = 0.010 \text{ m}, \ \Delta L = 5.0 \text{ mm} = 0.0050 \text{ m}$.
Assume a circular cross section: $A = \pi r^2 = \pi(\frac{D}{2})^2 = \pi(0.0050 \text{ m})^2 = 7.85 \times 10^{-5} \text{ m}^2$.
Solve:

$$F = \frac{YA}{L}\Delta L = \frac{(20 \times 10^{10} \text{ N/m}^2)(7.85 \times 10^{-5} \text{ m}^2)}{10 \text{ m}} 0.0050 \text{ m} = 7900 \text{ N}$$

This is the force required to stretch a steel cable of the given length and diameter by 5.0 mm.
Assess: A 1-cm-diameter cable is fairly substantial, so it ought to take a few thousand newtons to stretch it 5.0 mm. Notice the m^2 cancel in the numerator and so do the other m, leaving only N.

P8.33. Prepare: Equation 8.6 relates the quantities in question; the fractional decrease in length will be $\Delta L/L$, so rearrange the equation so $\Delta L/L$ is isolated.
Look up Young's modulus for Douglas fir in Table 8.1: $Y_{\text{Douglas fir}} = 1 \times 10^{10} \text{ N/m}^2$.
The total cross section will be three times the area of one leg:

$$A_{\text{tot}} = 3(\pi r^2) = 3\left(\pi\left(\frac{D}{2}\right)^2\right) = 3\pi(0.010 \text{ m})^2 = 9.42 \times 10^{-4} \text{ m}^2$$

Compute $F = w = mg = (75 \text{ kg})(9.8 \text{ m/s}^2) = 735 \text{ N}$.
Solve:

$$\frac{\Delta L}{L} = \frac{F}{AY} = \frac{735 \text{ N}}{(9.42 \times 10^{-4} \text{ m}^2)(1 \times 10^{10} \text{ N/m}^2)} = 7.8 \times 10^{-5}$$

This is a 0.0078% change in length.
Assess: We were not given the original length of the stool legs, but regardless of the original length, they decrease in length by only a small percentage—0.0078%—because F isn't large but A is.

P8.37. Prepare: Neglect the weight of the arm. The arm is $45°$ below horizontal which introduces $\sin 45°$ in each term, but it will cancel out.
Solve:

$$\Sigma\tau = T(14 \text{ cm})\sin 45° - (10 \text{ kg})(9.8 \text{ m/s}^2)(35 \text{ cm})\sin 45° = 0 \quad \Rightarrow \quad T = 860 \text{ N}$$

Assess: This answer is in the general range of the results in the example.

P8.39. Prepare: Assume equilibrium and compute the torques around the point labeled "pivot." The angle between the lever arm and the force is $60°$ for the weight of the torso and the head and arms.
 Solve:
(a)

$$\Sigma\tau = T\left(\frac{2}{3}L\right)\sin 12° - (320 \text{ N})\left(\frac{1}{2}L\right)\sin 60° - (160 \text{ N})(L)\sin 60° = 0$$

$$T = \frac{(320 \text{ N})(L)\sin 60°}{\frac{2}{3}L\sin 12°} = (480 \text{ N})\frac{\sin 60°}{\sin 12°} = 2000 \text{ N}$$

(b) Align the x-axis with the spine to find the force from the pelvic girdle.

$$\Sigma F_x = F_{\text{p.g.}} - (320 \text{ N})\cos 60° - (160 \text{ N})\cos 60° - (2000 \text{ N})\cos 12° = 0$$

$$F_{\text{p.g.}} = 2196 \text{ N} \approx 2200 \text{ N}$$

Assess: These are large forces, which is why they can cause back injuries. Squat when you lift.

P8.45. Prepare: Assume that the spring is ideal and it obeys Hooke's law. We also model the 5.0 kg mass as a particle. We will use the subscript s for the scale and sp for the spring. With the y-axis representing vertical positions, pictorial representations and free-body diagrams are shown for parts **(a)** through **(c)**.

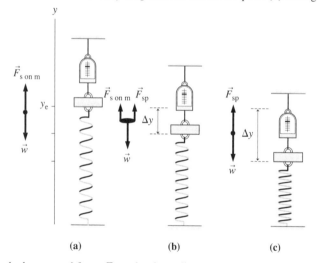

Solve: **(a)** The scale reads the upward force $F_{\text{s on m}}$ that it applies to the mass. Newton's second law gives

$$\Sigma(F_{\text{on m}})_y = F_{\text{s on m}} - w = 0 \Rightarrow F_{\text{s on m}} = w = mg = (5.0 \text{ kg})(9.8 \text{ m/s}^2) = 49 \text{ N}$$

(b) In this case, the force is

$$\Sigma(F_{\text{on m}})_y = F_{\text{s on m}} + F_{\text{sp}} - w = 0 \Rightarrow 20 \text{ N} + k\Delta y - mg = 0 \Rightarrow k = (mg - 20 \text{ N})/\Delta y$$
$$= (49 \text{ N} - 20 \text{ N})/0.02 \text{ m} = 1450 \text{ N/m}$$

(c) In this case, the force is

$$\Sigma(F_{\text{on m}})_y = F_{\text{sp}} - w = 0 \Rightarrow k\Delta y - mg = 0 \Rightarrow \Delta y = mg/k = (49 \text{ N})/(1450 \text{ N/m}) = 0.0338 \text{ m} = 3.4 \text{ cm}$$

P8.47. Prepare: Both springs are compressed by the same amount: $\Delta x = 1.00 \text{ cm}$. Each spring obeys Hooke's law (we assume we are in the linear region of the springs) and so exerts a force back on the block with a magnitude of $F_{\text{sp}} = k\Delta x$. The net spring force will simply be the sum of the two individual spring forces.

Solve:

$$(F_{sp})_1 = k_1 \Delta x = (12.0 \text{ N/cm})(1.00 \text{ cm}) = 12.0 \text{ N}$$

$$(F_{sp})_2 = k_2 \Delta x = (5.4 \text{ N/cm})(1.00 \text{ cm}) = 5.4 \text{ N}$$

$$(F_{sp})_{tot} = (F_{sp})_1 + (F_{sp})_2 = 12.0 \text{ N} + 5.4 \text{ N} = 17.4 \text{ N}$$

Assess: We have purposefully omitted the negative sign in Hooke's law since it only reflects the fact that the force and the stretch are in opposite directions—something we had kept in mind, but did not worry about since we only needed the magnitude of the forces.

These two springs are said to be in parallel, and they are equivalent to one spring whose spring constant is the sum of the spring constants of the two parallel springs.

P8.51. Prepare: We will model the student (S) as a particle and the spring as obeying Hooke's law. The only two forces acting on the student are his weight and the force due to the spring.

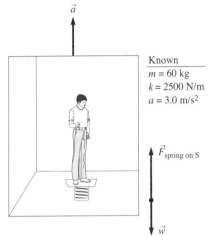

Known
$m = 60$ kg
$k = 2500$ N/m
$a = 3.0$ m/s^2

Solve: According to Newton's second law the force on the student is

$$\Sigma(F_{on\,S})_y = F_{spring\,on\,S} - w = ma_y \Rightarrow F_{spring\,on\,S} = w + ma_y = mg + ma_y = (60 \text{ kg})(9.8 \text{ m/s}^2 + 3.0 \text{ m/s}^2) = 768 \text{ N}$$

since $F_{spring\,on\,S} = F_{S\,on\,spring} = k\Delta y$, $k\Delta y = 768$ N. This means $\Delta y = (768 \text{ N})/(2500 \text{ N/m}) = 0.307 \text{ m} = 0.31 \text{ m}$.

P8.53. Prepare: There seem to be a lot of pieces to this puzzle; where does one start? Since the only horizontal force on the right block is due to the spring, let's start there: $F_{sp} = k\Delta x = (1000 \text{ N/m})(0.015 \text{ m}) = 15 \text{ N}$.

We apply Newton's second law to find the acceleration of the right block. $F_{net} = F_{sp} = ma$:

$$a = F_{sp}/m = 15 \text{ N}/3.0 \text{ kg} = 5.0 \text{ m/s}^2$$

For part **(b)** we note that since the spring's compression is constant over the time interval the two blocks must move together; i.e., they must have the same acceleration, so the left block's acceleration is also 5.0 m/s^2.

Also note that while the spring pushes to the right on the right block, it pushes to the left on the left block.

Solve: **(a)** We now apply Equation 2.11 (since the acceleration is constant) to find the final velocity of the right block at $t = 1.0$ s.

$$(v_x)_f = (v_x)_i + a_x \Delta t = 3.2 \text{ m/s} + (5.0 \text{ m/s}^2)(1.0\text{s}) = 8.2 \text{ m/s}$$

(b) Think about a simple free-body diagram for the left block and apply Newton's second law, noting that F_{sp} is subtracted because it is to the left.

$$F_{net} = F - F_{sp} = ma$$

$$F = F_{sp} + ma = 15 \text{ N} + (3.0 \text{ kg})(5.0 \text{ m/s}^2) = 30 \text{ N}$$

Assess: This is an interesting situation to analyze from a slightly different perspective. Think about both blocks making up a system (the parts of which are connected by the spring). It will take a certain amount of force F to accelerate that system at 5.0 m/s². Now isolate the right block and call it a new system. It has half the mass of the old system and so will need only half the force to accelerate it at the same rate. That is, F must accelerate twice as much mass as F_{sp} does on the right block.

P8.57. Prepare: Model the disk as a short wide rod. We are asked for the strain—the fractional change in length of the disk. We can solve for $\Delta L/L$ from Equation 8.7.

Assume that the disk is circular so that the area is $A = \pi R^2 = \pi (D/2)^2 = \pi (0.020 \text{ m})^2 = 0.00126 \text{ m}^2$.

The force is half the weight of the person: $F = \frac{1}{2}mg = \frac{1}{2}(65 \text{ kg})(9.8 \text{ m/s}^2) = 319 \text{ N}$.

Young's modulus for cartilage is not given in the chapter, but is in the problem: $Y = 1.0 \times 10^6 \text{ N/m}^2$.

Solve: Solve Equation 8.6 for $\Delta L/L$.

$$\frac{\Delta L}{L} = \frac{F}{YA} = \frac{319 \text{ N}}{(1.0 \times 10^{10} \text{ N/m}^2)(0.00126 \text{ m}^2)} = 0.000025 = 0.0025\%$$

Assess: This means the disk compresses by only a tiny amount. This seems reasonable. Notice that the actual thickness of the disk, given as 0.50 cm, is not needed in the calculation of the fractional compression.

P8.59. Prepare: We'll use the data from Example 8.10: $m_{\text{original}} = 70 \text{ kg}$ and $A_{\text{original}} = 4.8 \times 10^{-4} \text{ m}^2$.

The femur is not solid cortical bone material; we model it as a tube with an inner diameter and an outer diameter. Look up Young's modulus for cortical bone in Table 8.1.

Solve: **(a)** Both the inner and outer diameters are increased by a factor of 10; however, the cross-sectional area of the bone material does not increase by a factor of 10. Instead, because $A = \pi R^2$, the outer cross-sectional area and the inner cross-sectional area (the "hollow" of the tube) both increase by a factor of 100. But this means that the cross-sectional area of the bone material (the difference of the outer and inner areas) also increases by a factor of 100. So the new area is $A_{\text{new}} = 100(4.8 \times 10^{-4} \text{ m}^2) = 4.8 \times 10^{-2} \text{ m}^2$.

(b) Since volume is a three-dimensional concept, if we increase each linear dimension by a factor of 10 then the volume increases by a factor of $10^3 = 1000$. We assume the density of the man is the same as before, so his mass increases by the same factor as the volume: $m_{\text{new}} = 1000(70 \text{ kg}) = 70\,000 \text{ kg}$.

(c) We follow the strategy of Example 8.10. The force compressing the femur is the man's weight, $F = mg = (70,000 \text{ kg})(9.8 \text{ m/s}^2) = 690,000 \text{ N}$. The resulting stress on the femur is

$$\frac{F}{A} = \frac{690,000 \text{ N}}{4.8 \times 10^{-2} \text{ m}^2} = 1.4 \times 10^7 \text{ N/m}^2$$

A stress of 1.4×10^7 N/m² is 14% of the tensile strength of cortical bone given in Table 8.4.

Assess: This scaling problem illustrates clearly why animals of different sizes have different proportions. Because the volume scales with the cube of the linear dimensions and the area scales with the square of the linear dimensions then the force in F/A grows more quickly than the cross sectional area does.

PptI.21. Reason:
(a)

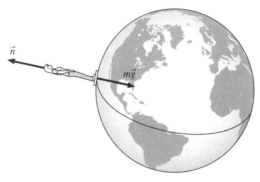

(b) The new reading is n, and the amount the reading is reduced is $mg - n$. The radius of the earth is 6.37×10^6 m. The mass of the person is $(800 \text{ N}) / (9.8 \text{ m/s}^2) = 81.6$ kg.   1d = 86400 s

$$mg - n = \Sigma F = ma = m \left(\frac{v^2}{r} \right) = m \frac{\left(\frac{2\pi r}{\Delta t} \right)^2}{r} = m \left(\frac{2\pi}{\Delta t} \right)^2 r =$$

$$(81.6 \text{ kg}) \left(\frac{2\pi}{86400 \text{ s}} \right)^2 (6.37 \times 10^6 \text{ m}) = 3.369 \text{ N} \approx 3.4 \text{ N}$$

Assess: The reading is only 3.4 N less than 800 N.

MOMENTUM

Q9.1. Reason: The velocities and masses vary from object to object, so there is no choice but to compute $p_x = mv_x$ for each one and then compare.

$$p_{1x} = (20 \text{ g})(1 \text{ m/s}) = 20 \text{ g} \cdot \text{m/s}$$
$$p_{2x} = (20 \text{ g})(2 \text{ m/s}) = 40 \text{ g} \cdot \text{m/s}$$
$$p_{3x} = (10 \text{ g})(2 \text{ m/s}) = 20 \text{ g} \cdot \text{m/s}$$
$$p_{4x} = (10 \text{ g})(1 \text{ m/s}) = 10 \text{ g} \cdot \text{m/s}$$
$$p_{5x} = (200 \text{g})(0.1 \text{m/s}) = 20 \text{g} \cdot \text{m/s}$$

So the answer is $p_{2x} > p_{1x} = p_{3x} = p_{5x} > p_{4x}$.

Assess: The largest, most massive object did not have the greatest momentum because it was moving slower than the rest.

Q9.3. Reason: When the question talks about forces, times, and momenta, we immediately think of the impulse-momentum theorem, which tells us that to change the momentum of an object we must exert a net external force on it over a time interval: $\vec{F}_{avg} \Delta t = \Delta \vec{p}$.

Because equal forces are exerted over equal times, the impulses are equal and the changes in momentum are equal. Because both carts start from rest, their changes in momentum are the same as the final momentum for each, so their final momenta are equal.

Assess: Notice that we did not need to know the mass of either cart, or even the specific time interval (as long as it was the same for both carts) to answer the question.

Q9.7. Reason: (a) Both particles cannot be at rest immediately after the collision. If they were both at rest, then the sum of the momenta after the collision would be zero, and since momentum is conserved in collisions, it would have had to be zero before as well (and it wasn't).

(b) If the masses are equal and the collision elastic, the moving particle will stop and give all of its momentum to the previously resting particle. A good example of this appears when a billiard ball rolls directly into another resting billiard ball.

Assess: We say momentum is conserved in all collisions because we assume that both colliding objects are part of the system and we assume the "impulse approximation" that other forces can be neglected during the short time interval of the collision.

In part **(a)** if the system contained a third particle that participated in the collision, then it is possible for the first two particles to end up at rest if the momentum is carried off by the third.

Q9.11. Reason: See Example 9.5. The two skaters interact with each other, but they form an isolated system because, for each skater, the upward normal force of the ice balances their downward weight force to make $\vec{F}_{net} = \vec{0}$. Thus the total momentum of the system of the two skaters will be conserved. Assume that both skaters are at rest before the push so that the total momentum before they push off is $\vec{P}_i = \vec{0}$. Consequently, the total momentum will still be $\vec{0}$ after they push off.

(a) Because the total momentum of the two-skater system is $\vec{0}$ after the push off, Megan and Jason each have momentum of the same magnitude but in the opposite direction as the other. Therefore the magnitude Δp is the same for each: $\vec{F}_{avg} \Delta t = \Delta \vec{p}$.

From the impulse-momentum theorem each experiences the same amount of impulse.

(b) They each experience the same amount of impulse because they experience the same magnitude force over the same time interval. However, over that time interval they do not experience the same acceleration. $\vec{F}_{net} = m\vec{a}$ says that since Megan and Jason experience the same force but Megan's mass is half of Jason's, then Megan's acceleration during push off will be twice Jason's. So she will have the greater speed at the end of the push off Δt.

Assess: It is important to think about both results until you are comfortable with them.

Q9.15. Reason: Assume that each angular momentum is to be calculated about the axis of symmetry. It will be useful to derive a general formula for the angular momentum of a particle in uniform circular motion of radius r, calculated around the axis of symmetry. Equation 7.14 reminds us that I for such a situation is mr^2, and Equation 6.7 tells us that $\omega = v/r$. Putting this all together gives the angular momentum for a particle in uniform circular motion: $L = I\omega = (mr^2)(v/r) = rmv$.

So we compute $L = rmv$ for each of the five situations.

$$L_1 = (2 \text{ m})(2 \text{ kg})(2 \text{ m/s}) = 8 \text{ kg} \cdot \text{m}^2/\text{s}$$
$$L_2 = (2 \text{ m})(3 \text{ kg})(1 \text{ m/s}) = 6 \text{ kg} \cdot \text{m}^2/\text{s}$$
$$L_3 = (2 \text{ m})(1 \text{ kg})(3 \text{ m/s}) = 6 \text{ kg} \cdot \text{m}^2/\text{s}$$
$$L_4 = (4 \text{ m})(2 \text{ kg})(1 \text{ m/s}) = 8 \text{ kg} \cdot \text{m}^2/\text{s}$$
$$L_5 = (4 \text{ m})(2 \text{ kg})(2 \text{ m/s}) = 16 \text{ kg} \cdot \text{m}^2/\text{s}$$

Finally, comparison gives $L_5 > L_1 = L_4 > L_2 = L_3$.

Assess: Since $p = mv$ the angular momentum for a particle in uniform circular motion can also be written $L = rp$, or, more generally, $L = rp_\perp$ (compare with Equation 7.3).

Q9.17. Reason: Since there is no net torque on the earth, the angular momentum of the earth is conserved. As water from the polar ice caps moves farther from the earth's rotation axis, the moment of inertia of the earth with increase from consideration of Equation 7.14. In order to keep the angular momentum of the earth constant, the angular velocity of the earth must decrease. If the angular velocity of the earth decreases, the period of rotation will increase, so the length of the day will increase.

Assess: Since the mass of the polar ice caps is very small compared to the mass of the entire earth, the effect on the length of the day will probably be small.

Problems

P9.3. Prepare: From Equations 9.5, 9.2, and 9.8, Newton's second law can be profitably rewritten as

$$\vec{F}_{avg} = \frac{\Delta \vec{p}}{\Delta t}$$

In fact, this is much closer to what Newton actually wrote than $\vec{F} = m\vec{a}$.

Solve: This allows us to find the force on the snowball. By Newton's third law we know that the snowball exerts a force of equal magnitude on the wall.

$$\vec{F}_{avg} = \frac{\Delta \vec{p}}{\Delta t} = \frac{m\vec{v}_f - m\vec{v}_i}{\Delta t} = \frac{m(\vec{v}_f - \vec{v}_i)}{\Delta t} = \frac{(0.12 \text{ kg})(0 \text{ m/s} - 7.5 \text{ m/s})}{0.15 \text{ s}} = -6.0 \text{ N}$$

where the negative sign indicates that the force on the snowball is opposite its original momentum. So the force on the wall is also 6.0 N.

Assess: This is not a large force, but the snowball has low mass, a moderate speed, and the collision time is fairly long.

P9.5. Prepare: We use the equation $J = \Delta p$.

Solve: The initial momentum of sled and rider is $p_i = mv_i = (80 \text{ kg})(4.0 \text{ m/s}) = 320 \text{ kg} \cdot \text{m/s}$ and the final momentum of sled and rider is $p_f = mv_f = (80 \text{ kg})(3.0 \text{ m/s}) = 240 \text{ kg} \cdot \text{m/s}$. So the impulse is given by

$$J = p_f - p_i = 320 \text{ kg} \cdot \text{m/s} - 240 \text{ kg} \cdot \text{m/s} = 80 \text{ kg} \cdot \text{m/s} = 80 \text{ N} \cdot \text{s}$$

Assess: This is a reasonable impulse. It could result, for example, from a force of around nine pounds for two seconds.

P9.7. Prepare: Please refer to Figure P9.7. Model the object as a particle and its interaction with the force as a collision. We will use Equations 9.1 and 9.9. Because $p = mv$, so $v = p/m$.

Solve: (a) Using the equations

$$(p_x)_f = (p_x)_i + J_x$$

$$J_x = \text{area under the force curve} \Rightarrow (v_x)_f = \left(1.0 \text{ m/s}\right) + \frac{1}{2.0 \text{ kg}}(\text{area under the force curve})$$

$$= (1.0 \text{ m/s}) + \frac{1}{2.0 \text{ kg}}(1.0 \text{ N s}) = 1.5 \text{ m/s}$$

(b) Likewise,

$$(v_x)_f = (1.0 \text{ m/s}) + \left(\frac{1}{2.0 \text{ kg}}\right) (\text{area under the force curve}) = (1.0 \text{ m/s}) + \left(\frac{1}{2.0 \text{ kg}}\right)(-1.0 \text{ N s}) = 0.5 \text{ m/s}$$

Assess: For an object with positive velocity, a negative impulse slows down an object and a positive impulse increases speed. The opposite is true for an object with negative velocity.

P9.9. Prepare: From Equations 9.5, 9.2, and 9.8, Newton's second law can be profitably rewritten as

$$\vec{F}_{avg} = \frac{\Delta \vec{p}}{\Delta t}$$

In fact, this is much closer to what Newton actually wrote than $\vec{F} = m\vec{a}$.

Solve: This allows us to find the force on the child and sled.

$$\vec{F}_{avg} = \frac{\Delta \vec{p}}{\Delta t} = \frac{m\vec{v}_f - m\vec{v}_i}{\Delta t} = \frac{m(\vec{v}_f - \vec{v}_i)}{\Delta t} = \frac{(35 \text{ kg})(0 \text{ m/s} - 1.5 \text{ m/s})}{0.50 \text{ s}} = -105 \text{ N} \approx -110 \text{ N}$$

where the negative sign indicates that the force is in the direction opposite the original motion, as stated in the problem. So the *amount* (magnitude) of the average force you need to exert is 105 N.

Assess: This result is neither too large nor too small. In some collision problems Δt is quite a bit shorter and so the force is correspondingly larger.

P9.13. Prepare: We'll call the system the two carts and consider it an isolated system so we can apply the law of conservation of momentum. The action all takes place in one dimension, so we don't need y-components. Let the subscript 1 stand for the small cart and 2 for the large.

Solve:

$$(P_x)_i = (P_x)_f$$

$$(p_{1x})_i + (p_{2x})_i = (p_{1x})_f + (p_{2x})_f$$

$$m_1(v_{1x})_i + m_2(v_{2x})_i = m_1(v_{1x})_f + m_2(v_{2x})_f$$

We want to know $(v_{2x})_f$ so we solve for it. Also recall that $(v_{2x})_i = 0 \text{ m/s}$, so the middle term in the following numerator drops out. The small cart recoils, which means its velocity after the collision is negative.

$$(v_{2x})_f = \frac{m_1(v_{1x})_i + m_2(v_{2x})_i - m_1(v_{1x})_f}{m_2} = \frac{(0.100 \text{ kg})(1.20 \text{ m/s}) - (0.100 \text{ kg})(-0.850 \text{ m/s})}{1.00 \text{ kg}} = 0.205 \text{ m/s}$$

Assess: The large cart does not move quickly, but the answer is reasonable because of the greater mass of the large cart.
We have followed the significant figure rules and kept three significant figures.

P9.17. Prepare: We will choose car + gravel to be our system. The initial *x*-velocity of the car is 2 m/s and that of the gravel is 0 m/s. To find the final *x*-velocity of the system, we will apply the momentum conservation Equation 9.15.

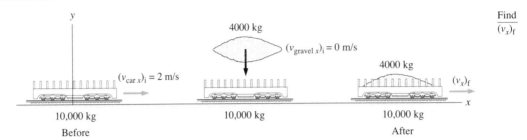

Solve: There are no *external* forces on the car + gravel system, so the horizontal momentum is conserved. This means $(p_x)_f = (p_x)_i$. Hence,

$$(10,000\ \text{kg} + 4,000\ \text{kg})(v_x)_f = (10,000\ \text{kg})(2.0\ \text{m/s}) + (4,000\ \text{kg})(0.0\ \text{m/s}) \Rightarrow (v_x)_f = 1.4\ \text{m/s}$$

Assess: The motion of railroad has to be on a level track for conservation of linear momentum to hold. As we would have expected, the final speed is smaller than the initial speed.

P9.19. Prepare: We will define our system to be archer + arrow. The force of the archer (A) on the arrow (a) is equal to the force of the arrow on the archer. These are internal forces within the system. The archer is standing on frictionless ice, and the normal force by ice on the system balances the weight force. Thus $\vec{F}_{\text{ext}} = \vec{0}$ on the system, and momentum is conserved. The initial momentum p_{ix} of the system is zero, because the archer and the arrow are at rest. The final moment p_{fx} must also be zero.

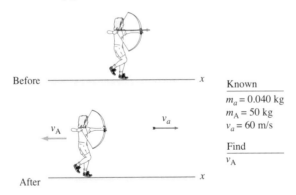

Solve: We have $M_A v_A + m_a v_a = 0$ kg m/s. Therefore,

$$v_A = \frac{-m_a v_a}{m_A} = \frac{-(0.04\ \text{kg})(60\ \text{m/s})}{50\ \text{kg}} = -0.48\ \text{m/s}$$

The archer's recoil *speed* is 0.48 m/s.
Assess: It is the total final momentum that is zero, although the individual momenta are nonzero. Since the arrow has forward momentum, the archer will have backward momentum.

P9.21. Prepare: We will define our system to be bird + bug. This is the case of an inelastic collision because the bird and bug move together after the collision. Horizontal momentum is conserved because there are no external forces acting on the system during the collision.

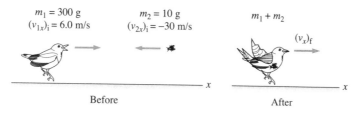

Solve: The conservation of momentum equation $p_{\text{f}x} = p_{\text{i}x}$ is

$$(m_1 + m_2)(v_x)_{\text{f}} = m_1(v_{1x})_{\text{i}} + m_2(v_{2x})_{\text{f}}$$
$$\Rightarrow (300\text{ g} + 10\text{ g})(v_x)_{\text{f}} = (300\text{ g})(6.0\text{ m/s}) + (10\text{ g})(-30\text{ m/s}) \Rightarrow (v_x)_{\text{f}} = 4.8\text{ m/s}$$

Assess: We left masses in grams, rather than convert to kilograms, because the mass units cancel out from both sides of the equation. Note that $(v_{2x})_{\text{i}}$ is negative. As would have been expected, the final speed is a little lower than the initial speed because (1) the bug has finite mass and (2) the bug has relatively large speed compared to the bird.

P9.23. Prepare: Even though this is an inelastic collision, momentum is still conserved during the short collision if we choose the system to be spitball plus carton. Let SB stand for the spitball, CTN the carton, and BOTH be the combined object after impact (we assume the spitball sticks to the carton). We are given $m_{\text{SB}} = 0.0030$ kg, $m_{\text{CTN}} = 0.020$ kg, and $(v_{\text{BOTH}x})_{\text{f}} = 0.30$ m/s.
Solve:

$$(P_x)_{\text{i}} = (P_x)_{\text{f}}$$
$$(p_{\text{SB}x})_{\text{i}} + (p_{\text{CTN}x})_{\text{i}} = (p_{\text{BOTH}x})_{\text{f}}$$
$$m_{\text{SB}}(v_{\text{SB}x})_{\text{i}} + m_{\text{CTN}}(v_{\text{CTN}x})_{\text{i}} = (m_{\text{SB}} + m_{\text{CTN}})(v_{\text{BOTH}x})_{\text{f}}$$

We want to know $(v_{\text{SB}x})_{\text{i}}$ so we solve for it. Also recall that $(v_{\text{CTN}x})_{\text{i}} = 0$ m/s so the last term in the following numerator drops out.

$$(v_{\text{SB}x})_{\text{i}} = \frac{(m_{\text{SB}} + m_{\text{CTN}})(v_{\text{BOTH}x})_{\text{f}} - m_{\text{CTN}}(v_{\text{CTN}x})_{\text{i}}}{m_{\text{SB}}} = \frac{(0.0030\text{ kg} + 0.020\text{ kg})(0.30\text{ m/s})}{0.0030\text{ kg}} = 2.3\text{ m/s}$$

Assess: The answer of 2.3 m/s is certainly within the capability of an expert spitballer.

P9.27. Prepare: We assume that the momentum is conserved in the collision. Please refer to Figure P9.27.
Solve: The conservation of momentum Equation 9.14 yields

$$(p_{1x})_{\text{f}} + (p_{2x})_{\text{f}} = (p_{1x})_{\text{i}} + (p_{2x})_{\text{i}} \Rightarrow (p_{1x})_{\text{f}} + 0\text{ kg}\cdot\text{m/s} = 2\text{ kg}\cdot\text{m/s} - 4\text{ kg}\cdot\text{m/s} \Rightarrow (p_{1x})_{\text{f}} = -2\text{ kg}\cdot\text{m/s}$$
$$(p_{1y})_{\text{f}} + (p_{2y})_{\text{f}} = (p_{1y})_{\text{i}} + (p_{2y})_{\text{i}} \Rightarrow (p_{1y})_{\text{f}} - 1\text{ kg}\cdot\text{m/s} = 2\text{ kg}\cdot\text{m/s} + 1\text{ kg}\cdot\text{m/s} \Rightarrow (p_{1y})_{\text{f}} = 4\text{ kg}\cdot\text{m/s}$$

The final momentum vector of particle 1 that has the above components is shown below.

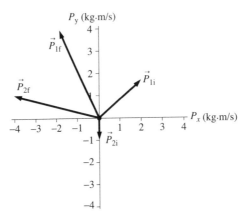

P9.29. Prepare: This problem deals with a case that is the opposite of a collision. Our system is comprised of three coconut pieces that are modeled as particles. During the blow up or "explosion," the total momentum of the system is conserved in the x-direction and the y-direction. We can thus apply Equation 9.14.

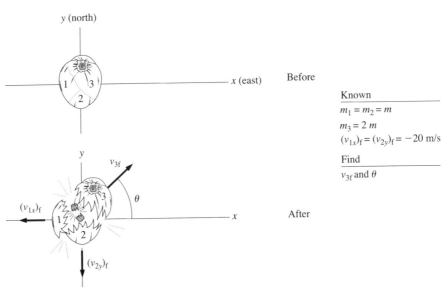

Known

$m_1 = m_2 = m$
$m_3 = 2\,m$
$(v_{1x})_f = (v_{2y})_f = -20$ m/s

Find

v_{3f} and θ

Solve: The initial momentum is zero. From $(p_x)_f = (p_x)_i$ we get

$$+m_1(v_{1x})_f + m_3(v_{3f})\cos\theta = 0\ \text{kg m/s} \Rightarrow (v_{3f})\cos\theta = \frac{-m_1(v_{fx})_1}{m_3} = \frac{-m(-20\ \text{m/s})}{2m} = 10\ \text{m/s}$$

From $(p_y)_f = (p_y)_i$, we get

$$+m_2(v_{2y})_f + m_3(v_{3f})\sin\theta = 0\ \text{kg m/s} \Rightarrow (v_{3f})\sin\theta = \frac{-m_2(v_{fy})_2}{m_3} = \frac{-m(-20\ \text{m/s})}{2m} = 10\ \text{m/s}$$

$$\Rightarrow (v_{3f}) = \sqrt{(10\ \text{m/s})^2 + (10\ \text{m/s})^2} = 14\ \text{m/s} \qquad \theta = \tan^{-1}(1) = 45°$$

Assess: The obtained speed of the third piece is of similar order of magnitude as the other two pieces, which is physically reasonable.

P9.31. Prepare: We will model the mother as a particle with $m = 47$ kg, $v = 4.2$ m/s, and $r = 2.6$ m. It will be useful to derive a general formula for the angular momentum of a particle in uniform circular motion of radius r, calculated around the axis of symmetry. Equation 7.14 reminds us that I for such a situation is mr^2, and Equation 6.7 tells us that $\left(-55 \text{ m/s}\right)\cos\left(25\right) = -23.2$ m/s. Putting this all together gives the angular momentum for a particle in uniform circular motion:

$$L = I\omega = (mr^2)(v/r) = rmv$$

Solve:

$$L = rmv = (2.6 \text{ m})(47 \text{ kg})(4.2 \text{ m/s}) = 510 \text{ kg} \cdot \text{m}^2/\text{s}$$

to two significant figures.

Assess: Until one gets a feel for how much angular momentum objects have it is difficult to know if our answer is reasonable, but the derivation of $L = rmv$ is straightforward and we checked our multiplication twice, so it is probably correct.

P9.33. Prepare: The disk is a rotating rigid body. Please refer to Figure P9.33. The angular velocity ω is 600 rpm $= 600 \times 2\pi/60$ rad/s $= 20\pi$ rad/s. From Table 7.4, the moment of inertial of the disk about its center is $(1/2)$ MR^2, which can be used with $L = I\omega$ to find the angular momentum.
Solve:

$$I = \frac{1}{2} MR^2 = \frac{1}{2}(2.0 \text{ kg})(0.020 \text{ m})^2 = 4.0 \times 10^{-4} \text{ kg} \cdot \text{m}^2$$

Thus, $L = I\omega = (4.0 \times 10^{-4} \text{ kg m}^2)(20\pi \text{ rad/s}) = 0.025 \text{ kg m}^2/\text{s}$. If we wrap our right fingers in the direction of the disk's rotation, our thumb will point in the $-x$ direction. Consequently,

$$\vec{L} = (0.025 \text{ kg} \cdot \text{m}^2/\text{s, into page})$$

Assess: Don't forget the direction of the angular momentum because it is a vector quantity.

P9.35. Prepare: We neglect any small frictional torque the ice may exert on the skater and apply the law of conservation of angular momentum.

$$L_i = L_f$$
$$I_i\omega_i = I_f\omega_f$$

Even though the data for $\omega_i = 5.0$ rev/s is not in SI units, it's okay because we are asked for the answer in the same units. We are also given $I_i = 0.80 \text{ kg} \cdot \text{m}^2$ and $I_f = 3.2 \text{ kg} \cdot \text{m}^2$.
Solve:

$$\omega_f = \frac{I_i\omega_i}{I_f} = \frac{(0.80 \text{ kg} \cdot \text{m}^2)(5.0 \text{ rev/s})}{3.2 \text{ kg} \cdot \text{m}^2} = 1.25 \text{ rev/s} \approx 1.3 \text{ rev/s}$$

Assess: I increased by a factor of 4, so we expect ω to decrease by a factor of 4.

P9.37. Prepare: Please refer to Figure P9.37. Model the glider cart as a particle, and its interaction with the spring as a collision. The initial and final speeds of the glider are shown on the velocity graph and the mass of the glider is known. We can thus find the momentum change of the glider, which is equal to the impulse. Impulse is also given by the area under the force graph, which we will find from the force graph in terms of Δt.

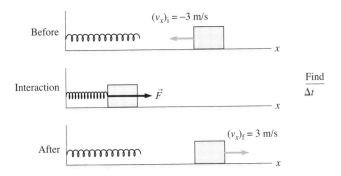

Solve: Using the impulse-momentum theorem $(p_x)_f - (p_x)_i = J_x$,

$$(0.6 \text{ kg})(3 \text{ m/s}) - (0.6 \text{ kg})(-3 \text{ m/s}) = \text{area under force curve} = \tfrac{1}{2}(36 \text{ N})(\Delta t) \Rightarrow \Delta t = 0.20 \text{ s}$$

Assess: You can solve this problem using kinematics to check your answer. From the graph you have the average force during compression to be 18 N, and therefore the average acceleration to be 30 m/s². Now calculate the time taken for the velocity to go from 3 m/s to 0 m/s, and twice this time should match the 0.20 s found in the previous figure.

P9.41. Prepare: We combine Equation 9.1 $J = F_{avg} \Delta t$ with the impulse-momentum theorem in the *y*-direction $J_y = \Delta p_y$. See Example 9.1. This tells us that Δp_y is the area under the curve of the net force in the vertical direction vs. time. And if we know the change in the woman's vertical momentum we can figure out the speed with which she leaves the ground; to do this last step, however, we'll need her mass.

Look at the first part of the graph while the force exerted by the floor is constant. During that time she isn't accelerating, so the force the floor exerts must be equal in magnitude to her weight; so she weighs 600 N and her mass is 600 N/(9.8 m/s²) = 61 kg.

It should also be clear from the graph that she leaves the floor at $t = 0.5$ s when the force of the floor on her is zero. The graph we are given is not the graph of the net force. The hint warns us that the upward force of the floor is not the only force on the woman. We have just concluded that the earth is exerting a downward gravitational force of 600 N (her weight) on her. Therefore a graph of the *net* vertical force on her vs. time would simply be the same graph only 600 N lower on the force axis.

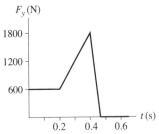

Note carefully that the graph now crosses the *t*-axis at $t = 0.475$s.

Solve: What is the area under the new graph? We'll take the area of the triangle above the *t*-axis to be positive and then subtract the area of the smaller triangle below the *t*-axis. The general formula for the area of a triangle is

$$A = \tfrac{1}{2} \times \text{height} \times \text{base}$$
$$A_{big} = \tfrac{1}{2}(1800 \text{ N})(0.275 \text{ s}) = 248 \text{ N} \cdot \text{s}$$
$$A_{small} = \tfrac{1}{2}(600 \text{ N})(0.025 \text{ s}) = 15 \text{ N} \cdot \text{s}$$

The total area (with the triangle above the axis positive and the triangle below the axis negative) is 248 N·s 15 N·s = 233 N·s; this is the vertical impulse on the woman, and is also equal to her change in vertical momentum. Since $\Delta p_y = m\Delta v_y$ we simply divide by her mass to find her change in velocity:

$$\Delta v_y = \Delta p_y / m = (233 \text{ N} \cdot \text{s})/(61 \text{ kg}) = 3.8 \text{ m/s}$$

Because she started from rest this value is also her final speed, just as she leaves the ground.

Assess: We assumed the ability to read the data from the graph to two significant figures. If we were not confident in this we would report the result to just one significant figure: $(v_y)_f \approx 4$ m/s. The answer of ≈ 4m/s does seem to be in the reasonable range.

It is worth following the units in the last equation to see the answer end up in m/s.

P9.47. Prepare: Let the system be ball + racket. During the collision of the ball and the racket, momentum is conserved because all external interactions are insignificantly small. We will also use the momentum-impulse theorem.

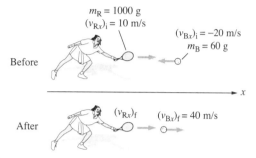

$m_R = 1000$ g
$(v_{Rx})_i = 10$ m/s
$(v_{Bx})_i = -20$ m/s
$m_B = 60$ g

Before

$(v_{Rx})_f$ $(v_{Bx})_f = 40$ m/s

After

Solve: (a) The conservation of momentum equation $(p_x)_f = (p_x)_i$ is

$$m_R(v_{Rx})_f + m_B(v_{Bx})_f = m_R(v_{Rx})_i + m_B(v_{Bx})_i$$

$$(1.0 \text{ kg})(v_{Rx})_f + (0.06 \text{ kg})(40 \text{ m/s}) = (1.0 \text{ kg})(10 \text{ m/s}) + (0.06 \text{ kg})(-20 \text{ m/s}) \Rightarrow (v_{Rx})_f = 6.4 \text{ m/s}$$

(b) The impulse on the ball is calculated from $(p_{Bx})_f = (p_{Bx})_i + J_x$ as follows:

$$(0.06 \text{ kg})(40 \text{ m/s}) = (0.06 \text{ kg})(-20 \text{ m/s}) + J_x \Rightarrow J_x = 3.6 \text{ N s} = F_{avg}\Delta t$$

$$\Rightarrow F_{avg} = \frac{3.6 \text{ Ns}}{10 \text{ ms}} = 360 \text{ N}$$

Assess: Let us now compare this force with the ball's weight $w_B = m_B g = (0.06 \text{ kg})(9.8 \text{ m/s}^2) = 0.588$ N. Thus, $F_{avg} = 610 \, w_B$. This is a significant force and is reasonable because the impulse due to this force changes the direction as well as the speed of the ball from approximately 45 mph to 90 mph.

P9.49. Prepare: We will define our system to be Dan + skateboard. The system has nonzero initial momentum p_{ix}. As Dan (D) jumps backward off the gliding skateboard (S), the skateboard will move forward in such a way that the final total momentum of the system is equal to the initial momentum. This conservation of momentum occurs because $\vec{F}_{ext} = \vec{0}$ on the system.

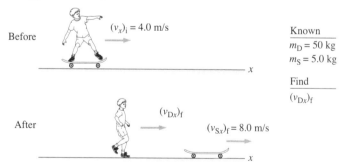

Before $(v_x)_i = 4.0$ m/s

Known
$m_D = 50$ kg
$m_S = 5.0$ kg

Find
$(v_{Dx})_f$

$(v_{Dx})_f$

After $(v_{Sx})_f = 8.0$ m/s

Solve: We have $m_S(v_{Sx})_f + m_D(v_{Dx})_f = (m_S + m_D)(v_x)_i$. Hence,

$$(5.0 \text{ kg})(8.0 \text{ m/s}) + (50 \text{ kg})(v_{Dx})_f = (5.0 \text{ kg} + 50 \text{ kg})(4.0 \text{ m/s}) \Rightarrow (v_{Dx})_f = 3.6 \text{ m/s}$$

Assess: A speed of 3.6 m/s or 8 mph is reasonable.

P9.53. Prepare: Model the train cars as particles. Since the train cars stick together, we are dealing with perfectly inelastic collisions. Momentum is conserved in the collisions of this problem.

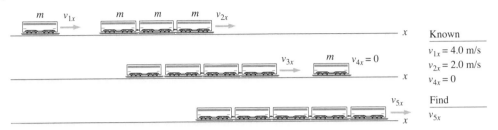

Solve: In the collision between the three-car train and the single car:

$$mv_{1x} + (3m)v_{2x} = 4mv_{3x} \Rightarrow v_{1x} + 3v_{2x} = 4v_{3x} \Rightarrow (4.0 \text{ m/s}) + 3(2.0 \text{ m/s}) = 4v_{3x} \Rightarrow v_{3x} = 2.5 \text{ m/s}$$

In the collision between the four-car train and the stationary car:

$$(4m)v_{3x} + mv_{4x} = (5m)v_{5x} \Rightarrow 4v_{3x} + 0 \text{ m/s} = 5v_{5x} \Rightarrow v_{5x} = \frac{4v_{3x}}{5} = (0.8)(2.5 \text{ m/s}) = 2.0 \text{ m/s}$$

Assess: The motion of railroad has to be on a level track for linear momentum to be constant. The speed of the five-car train, as expected, is of the same order of magnitude as the other speeds.

P9.59. Prepare: This is a two part problem. First, we have an inelastic collision between the wood block and the bullet. The bullet and the wood block are an isolated system. Since any external force acting during the collision is not going to be significant, the momentum of the system will be conserved. The second part involves the dynamics of the block + bullet sliding on the wood table. We treat the block and the bullet as particles.

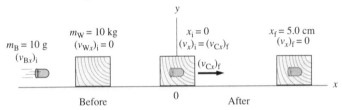

Solve: The equation $(p_x)_f = (p_x)_i$ gives

$$(m_B + m_W)(v_{Cx})_f = m_B(v_{Bx})_i + m_W(v_{Wx})_i$$

$$\Rightarrow (0.01 \text{ kg} + 10 \text{ kg})(v_{Cx})_f = (0.01 \text{ kg})(v_{Bx})_i + (10.0 \text{ kg})(0 \text{ m/s}) \Rightarrow (v_{Cx})_f = \frac{1}{1001}(v_{Bx})_i$$

Using $-f_k = -\mu_k n = -\mu_k(m_B + m_W)g = (m_B + m_W)a_x \Rightarrow a_x = -\mu_k g$. Note that the negative sign appears in front of f_k because the force of friction points in the $-x$ direction.

Using the kinematics equation $(v_x)_f^2 = (v_x)_i^2 + 2a_x(x_f - x_i)$,

$$0 = (v_{Cx})_f^2 - 2\mu_k g(x_f - x_i) \Rightarrow 0 \text{ m}^2/\text{s}^2 = \left(\frac{1}{1001}\right)^2 (v_{Bx})_i^2 - 2\mu_k g x_f \Rightarrow 0 \text{ m}^2/\text{s}^2 = \left(\frac{1}{1001}\right)^2 (v_{Bx})_i^2 - 2\mu_k g x_f$$

$$\Rightarrow (v_{Bx})_i = 1001\sqrt{2\mu_k g x_f} = 1001\sqrt{2(0.2)(9.8 \text{ m/s}^2)(0.05 \text{ m})} = 440 \text{ m/s}$$

Assess: The bullet's speed is reasonable (≈ 1000 mph).

P9.61. Prepare: We will model the two fragments of the rocket after the explosion as particles. We assume the explosion separates the two parts in a vertical manner. This is a three-part problem. In the first part, we will use kinematics equations to find the vertical position where the rocket breaks into two pieces. In the second part, we will apply conservation of momentum along the y direction to the system (that is, the two fragments) in the explosion. The momentum conservation "applies" because the forces involved during the explosion are much larger than the external force due to gravity during the small period that the explosion lasts.

In the third part, we will again use kinematics equations to find the velocity of the heavier fragment just after the explosion.

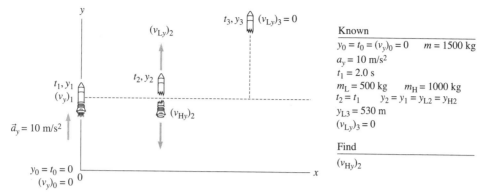

Solve: The rocket accelerates for 2.0 s from rest, so

$$(v_y)_1 = (v_y)_0 + a_y(t_1 - t_0) = 0 \text{ m/s} + (10 \text{ m/s}^2)(2 \text{ s} - 0 \text{ s}) = 20 \text{ m/s}$$

$$y_1 = y_0 + (v_y)_0(t_1 - t_0) + \frac{1}{2} a_y(t_1 - t_0)^2 = 0 \text{ m} + 0 \text{ m} + \frac{1}{2}(10 \text{ m/s}^2)(2 \text{ s})^2 = 20 \text{ m}$$

At the explosion the equation $(p_y)_f = (p_y)_i$ is

$$m_L(v_{Ly})_2 + m_H(v_{Hy})_2 = (m_L + m_H)(v_y)_1 \Rightarrow (500 \text{ kg})(v_{Ly})_2 + (1000 \text{ kg})(v_{Hy})_2 = (1500 \text{ kg})(20 \text{ m/s})$$

To find $(v_{Hy})_2$ we must first find $(v_{Ly})_2$, the velocity after the explosion of the upper section. Using kinematics,

$$(v_{Ly})_3^2 = (v_{Ly})_2^2 + 2(-9.8 \text{ m/s}^2)((y_L)_3 - (y_L)_2) \Rightarrow (v_{Ly})_2 = \sqrt{2(9.8 \text{ m/s}^2)(530 \text{ m} - 20 \text{ m})} = 99.98 \text{ m/s}$$

Now, going back to the momentum conservation equation we get

$$(500 \text{ kg})(99.98 \text{ m/s}) + (1000 \text{ kg})(v_{Hy})_2 = (1500 \text{ kg})(20 \text{ m/s}) \Rightarrow (v_{Hy})_2 = -20 \text{ m/s}$$

The negative sign indicates downward motion.

P9.63. Prepare: Choose the system to be cannon + ball. There are no significant external horizontal forces during the brief interval in which the cannon fires, so within the impulse approximation the horizontal momentum is conserved. We'll ignore the very small mass loss of the exploding gunpowder. The statement that the ball travels at 200 m/s relative to the cannon can be written $(v_{Bx})_f = (v_{Cx})_f + 200 \text{ m/s}$. That is, the ball's speed is 200 m/s more than the cannon's speed.

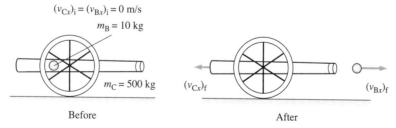

Solve: The initial momentum is zero, so the conservation of momentum equation $(p_x)_f = (p_x)_i$ is

$$(p_x)_f = m_C(v_{Cx})_f + m_B(v_{Bx})_f = m_C(v_{Cx})_f + m_B((v_{Cx})_f + 200 \text{ m/s}) = 0 \text{ kg m/s}$$

$$\Rightarrow (v_{Cx})_f = -\frac{m_B}{m_C + m_B}(200 \text{ m/s}) = -\frac{10 \text{ kg}}{510 \text{ kg}}(200 \text{ m/s}) = -3.9 \text{ m/s}$$

That is, the cannon recoils to the left (negative sign) at 3.9 m/s. Thus the cannonball's speed relative to the ground is $(v_{Bx})_f = -3.9$ m/s + 200 m/s = 196 m/s.

Assess: An expected relatively small recoil speed is essentially due to the large mass of the cannon.

P9.69. Prepare: Because there's no friction or other tangential forces, the angular momentum of the block + rod system is conserved. The rod is massless, so the angular momentum is entirely that of a mass in circular motion.

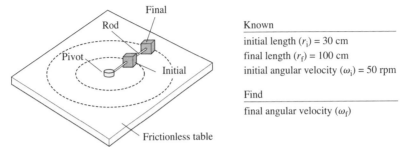

Known
initial length (r_i) = 30 cm
final length (r_f) = 100 cm
initial angular velocity (ω_i) = 50 rpm

Find
final angular velocity (ω_f)

Solve: The conservation of angular momentum equation $L_f = L_i$ is

$$mv_f r_f = mv_i r_i \Rightarrow (r_f \omega_f) r_f = (r_i \omega_i) r_i \Rightarrow \omega_f = \left(\frac{r_i}{r_f} \right)^2 \omega_i = \left(\frac{30 \text{ cm}}{100 \text{ cm}} \right)^2 (50 \text{ rpm}) = 4.5 \text{ rpm}$$

Assess: An angular speed of 4.5 rpm is reasonable, since the angular speed varies inversely as the square of the object's distance from the rotation axis.

P9.73. Prepare: Since there are no external torques, we can use the conservation of angular momentum to solve this problem. We need to use the moment of inertia of a rotating disk $I_{disk} = \frac{1}{2} MR^2$ in order to write the final angular momentum of the merry-go-round: $(L_m)_f = I_{disk} \omega_f$. The final angular momentum of Joey is given by $(L_J)_f = m(v_J)_f R$.

Solve: Before Joey begins running, both he and the merry-go-round are at rest, so the total angular momentum is 0. Let us say that he runs counterclockwise so that his final angular momentum is positive. Then the merry-go-round must rotate clockwise so that its final angular momentum is negative. In this way the angular momenta of Joey and the merry-go-round will still add to 0. The equation for conservation of angular momentum is

$$0 \text{ kg} \cdot \text{m}^2 / \text{s} = m(v_J)_f R + \left(\frac{1}{2} MR^2 \right) \omega_f \Rightarrow$$

$$0 \text{ kg} \cdot \text{m}^2 / \text{s} = (36 \text{ kg})(5.0 \text{ m/s})(2.0 \text{ m}) + \left(\frac{1}{2} \right)(200 \text{ kg})(2.0 \text{ m})^2 \omega_f$$

This equation can be solved for the angular velocity: $\omega_f = -0.90$ rad/s.

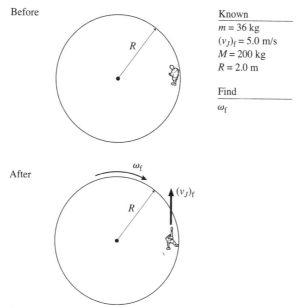

Before

Known
$m = 36$ kg
$(v_J)_f = 5.0$ m/s
$M = 200$ kg
$R = 2.0$ m

Find

ω_f

After

ω_f

$(v_J)_f$

R

Assess: The negative sign is as expected considering that the total angular momentum is zero. For this to be true, the two bodies, the merry-go-round and Joey, must move in opposite directions around the axis.

P9.75. Prepare: Define the system to be Disk A plus Disk B so that during the time of the collision there are no net torques on this isolated system. This allows us to use the law of conservation of angular momentum. After the collision the combined object will have a moment of inertia equal to the sum of the moments of inertia of Disk A and Disk B and it will have one common angular speed ω_f (which we seek).

We must remember to call counterclockwise angular speeds positive and clockwise angular speeds negative.

Known

$M_A = 2.0$ kg
$R_A = 0.40$ m
$(\omega_A)_i = -30$ rev/s
$M_B = 2.0$ kg
$R_B = 0.20$ m
$(\omega_B)_i = 30$ rev/s

Find

ω_f

Preliminarily compute the moments of inertia of the disks:

$$I_A = \frac{1}{2} M_A R_A^2 = \frac{1}{2}(2.0 \text{ kg})(0.40 \text{ m})^2 = 0.16 \text{ kg} \cdot \text{m}^2$$

$$I_B = \frac{1}{2} M_B R_B^2 = \frac{1}{2}(2.0 \text{ kg})(0.20 \text{ m})^2 = 0.040 \text{ kg} \cdot \text{m}^2$$

Solve:

$$\Sigma L_i = \Sigma L_f$$
$$(L_A)_i + (L_B)_i = (L_{A+B})_f$$
$$I_A(\omega_A)_i + I_B(\omega_B)_i = (I_A + I_B)\omega_f$$

Solving for ω_f gives

$$\omega_f = \frac{I_A(\omega_A)_i + I_B(\omega_B)_i}{I_A + I_B} = \frac{(0.16\,\text{kg}\cdot\text{m}^2)(-30\,\text{rev/s}) + (0.040\,\text{kg}\cdot\text{m}^2)(30\,\text{rev/s})}{0.16\,\text{kg}\cdot\text{m}^2 + 0.040\,\text{kg}\cdot\text{m}^2} = -18\,\text{rev/s}$$

That is, the angular speed is 18 rev/s and the direction is clockwise.

Assess: Since the two disks were rotating in opposite directions at the same speed we expect the angular speed afterwards to be less than the original speeds, and indeed it is.

ENERGY AND WORK

Q10.1. Reason: The brakes in a car slow down the car by converting its kinetic energy to thermal energy in the brake shoes through friction. Cars have large kinetic energies, and all of that energy is converted to thermal energy in the brake shoes, which causes their temperature to increase greatly. Therefore they must be made of material that can tolerate very high temperatures without being damaged.
Assess: This is an example of an energy conversion. All of the car's kinetic energy is converted to thermal energy through friction. To get an appreciation of how much kinetic energy is absorbed by the brake shoes, consider instead the energy explicit in stopping the car by hitting a stationary object instead!

Q10.7. Reason: Here we need to convert potential energy to kinetic energy without any work done on the system. Consider dropping a ball from a height. The ball's gravitational energy is converted to the kinetic energy of the ball as it falls. Another example would be releasing a ball at the end of a compressed spring. The potential energy stored in the compressed spring is converted to the kinetic energy of the ball as the spring stretches to its equilibrium length. Since no external forces act on a system, the work on the system is zero.
Assess: Many examples in the problem section will involve just this type of conversion of potential energy to kinetic energy. If no forces from the environment act on a system, the work done on the system is zero.

Q10.9. Reason: We need a process that converts potential energy totally into thermal energy without changing the kinetic energy. Consider a wood block sliding down a rough inclined surface at a constant speed. The gravitational potential energy is decreasing and the kinetic energy is constant. All the decrease in gravitational potential energy becomes an increase in thermal energy.
Assess: Gravitational potential energy decreases because there is a change in the height of the block. The kinetic energy does not change because the speed of the block is constant.

Q10.13. Reason: (a) The work done is $W = Fd$. Both particles experience the same force, so the greater work is done on the particle that undergoes the greater displacement. Particle A, which is less massive than B, will have the greater acceleration and thus travel further during the 1 s interval. Thus more work is done on particle A. **(b)** Impulse is $F\Delta t$. Both particles experience the same force F for the same time interval $\Delta t = 1$ s. Thus the same impulse is delivered to both particles. **(c)** Both particles receive the same impulse, so the change in their momenta is the same, that is, $m_A\left(v_f\right)_A = m_B\left(v_f\right)_B$. But because $m_A < m_B$, it must be that $\left(v_f\right)_A > \left(v_f\right)_B$. This result can also be found from kinematics, as in part (a).
Assess: Work is the product of the force and the displacement, while impulse is the product of the force and the time during which the force acts.

Q10.15. Reason: Neglecting frictional losses, the work you do on the jack is converted into gravitational potential energy of the car as it is raised. The work you do is Fd, where F is the force you apply to the jack handle and d is the 20 cm distance you move the handle. This work goes into increasing the potential energy by an amount $mgh = wh$, where w is the car's weight and $h = 0.2$ cm is the change in the car's height. So $Fd = wh$ so that $F / w = h / d$.
Assess: Because the force F you can apply is so much less than the weight w of the car, h must be much less than d.

Q10.19. Reason: Because both rocks are thrown from the same height, they have the same potential energy. And since they are thrown with the same speed, they have the same kinetic energy. Thus both rocks have the same total energy. When they reach the ground, they will have this same total energy. Because they're both at the same height at ground level, their potential energy there is the same. Thus they must have the same kinetic energy, and hence the same speed.

Assess: Although Chris's rock was thrown angled upward so that it slows as it first rises, it then speeds up as it begins to fall, attaining the same speed as Sandy's as it passes the initial height. Sandy's rock will hit the ground *first*, but its speed will be no greater than Chris's.

Q10.21. Reason: As you land, the force of the ground or pad does negative work on your body, transferring out the kinetic energy you have just before impact. This work is $-Fd$, where d is the distance over which your body stops. With the short stopping distance involved upon hitting the ground, the force F will be much greater than it is with the long stopping distance upon hitting the pad.

Assess: For a given amount of work, the force is large when the displacement is small.

Problems

P10.1. Prepare: Since definition this is an etiquette class and you are walking slowly and steadily, assume the book remains level. We will use the of work, Equation 10.6, to explicitly calculate the work done. Since no component of the force is along the displacement of the book, we expect the work done by your head will be zero.

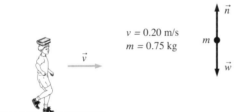

Solve: Refer to the diagram. There is no force in the horizontal direction since your velocity is constant. Your head exerts a normal force on the book, which counteracts the weight of the book. Since you are walking steadily there is no acceleration in the vertical direction, so the normal force is equal to the weight of the book. The force your head exerts on the book is then

$$n = w = (0.75 \text{ kg})(9.80 \text{ m/s}^2) = 7.4 \text{ N}$$

The work done on the book by your head is

$$W = Fd\cos\theta = (7.4 \text{ N})(2.5 \text{ m})\cos(90°) = (7.4 \text{ N})(2.5 \text{ m})(0) = 0.0 \text{ J}$$

The work done by your head on the book is exactly 0 Joules.

Assess: As expected, no work is done since the force and the displacement are at right angles. Note that your speed, which is given in the problem statement, is irrelevant.

P10.3. Prepare: Note that not all the forces in this problem are parallel to the displacement. Equation 10.6 gives the work done by a constant force which is not parallel to the displacement: $W = Fd\cos(\theta)$, where W is the work done by the force F at an angle θ to the displacement d. Here the displacement is exactly downwards in the same direction as $\vec{w}$. We will take all forces as having four significant figures (as implied by $T_2 = 1295$ N).

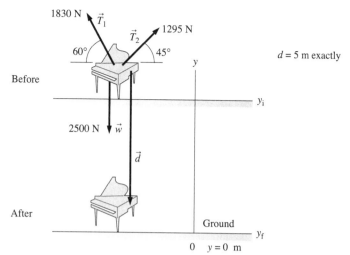

d = 5 m exactly

Solve: Refer to the diagram. The angle between the force $\vec{w}$ and the displacement is 0°, so

$$W_{\vec{w}} = wd\cos\theta = (2500\text{ N})(5\text{ m})\cos(0°) = 12.5\text{ kJ}$$

The angle between the force $\vec{T}_1$ and the displacement is $90° + 60° = 150°$.

$$W_{\vec{T}_1} = T_1 d\cos\theta = (1830\text{ N})(5\text{ m})\cos(150°) = -7.92\text{ kJ}$$

The angle between the tension $\vec{T}_2$ and the displacement is $90° + 45° = 135°$.

$$W_{\vec{T}_2} = T_2 d\cos\theta = (1295\text{ N})(5\text{ m})\cos(135°) = -4.58\text{ kJ}$$

Assess: Note that the displacement d in all these cases is directed downwards and that it is always the angle between the force and displacement used in the work equation. For example, the angle between $\vec{T}_1$ and $\vec{d}$ is 150°, not 60°.

P10.7. Prepare: The kinetic energy for any object moving of mass m with velocity v is given in Equation 10.8: $K = \frac{1}{2}mv^2$.

$v_B = 500$ m/s
$m_B = 10$ g

$v_{BB} = 10$ m/s
$m_{BB} = 10$ kg

Solve: For the bullet,

$$K_B = \frac{1}{2}m_B v_B^2 = \frac{1}{2}(0.010\text{ kg})(500\text{ m/s})^2 = 1.3\text{ kJ}$$

For the bowling ball,

$$K_{BB} = \frac{1}{2}m_{BB} v_{BB}^2 = \frac{1}{2}(10\text{ kg})(10\text{ m/s})^2 = 0.50\text{ kJ}$$

Thus, the bullet has the larger kinetic energy.
Assess: Kinetic energy depends not only on mass but also on the square of the velocity. The previous calculation shows this dependence. Although the mass of the bullet is 1000 times smaller than the mass of the bowling ball, its speed is 50 times larger, which leads to the bullet having over twice the kinetic energy of the bowling ball.

P10.9. Prepare: In order to work this problem, we need to know that the kinetic energy of an object is given by $K = mv^2 / 2$.

Solve: The problem may be solved in a qualitative manner or in a quantitative manner. Since some students think one way and some the other, we will use both methods.

(a) First, in a qualitative manner. Since the kinetic energy depends on the square of the speed, the kinetic energy will be doubled if the speed is increases by a factor of $\sqrt{2}$. This is true because $(\sqrt{2})^2 = 2$. Then the new speed is $\sqrt{2}(10 \text{ m}/\text{s}) = 14 \text{ m}/\text{s}$.

Second, in a more quantitive manner. Use a subscript 1 for the present case where the speed is 10 m/s and a subscript 2 for the new case where the speed is such that the kinetic energy is doubled.

$$K_1 = mv_1^2 / 2 \quad \text{and} \quad K_2 = mv_2^2 / 2$$

We want

$$K_2 = 2K_1$$

Inserting expressions for K_1 and K_2 obtain

$$\frac{mv_2^2}{2} = 2\frac{mv_1^2}{2} \quad \text{or} \quad v_2 = \sqrt{2}v_1 = 14 \text{ m}/\text{s}$$

(b) First, in a qualitative manner, if the speed is doubled and the kinetic energy depends on the square of the speed, the kinetic energy will increase by a factor of four.

Second, in a more quantitative manner, use a subscript 1 for the present case and a subscript 2 for the new case where the speed is doubled.

$$K_1 = mv_1^2 / 2 \quad \text{and} \quad K_2 = mv_2^2 / 2$$

We want $v_2 = 2v_1$. Inserting v_2 into K_2, we obtain

$$K_2 = \frac{mv_2^2}{2} = \frac{m(2v_1)^2}{2} = 4\left(\frac{mv_1^2}{2}\right) = 4K_1$$

This expression clearly shows that the kinetic energy is increased by a factor of four when the speed is doubled.

Assess: The key to the problem is to know that the kinetic energy depends on the speed squared. After that, we can approach the problem in a qualitative or a quantitative manner. You may prefer one method over the other, but you should be able to work in either mode.

P10.11. Prepare: Use the law of conservation of energy, Equation 10.4, to find the work done on the particle. We will assume there is no change in thermal energy of the ball.

Solve: Consider the system to be the plastic ball. Since there is no change in potential, thermal or chemical energy of the ball, and there is no heat leaving or entering the system, the conservation of energy equation becomes

$$W = \Delta K = \frac{1}{2}mv_f^2 - \frac{1}{2}mv_i^2 = \frac{1}{2}m\left(v_f^2 - v_i^2\right) = \frac{1}{2}(0.020 \text{ kg})[(30 \text{ m/s})^2 - (-30 \text{ m/s})^2] = 0 \text{ J}$$

Assess: Note that no work is done on the ball in reversing its velocity. This is because negative work is done in slowing the ball down to rest, and an equal amount of positive work is done in bringing the ball to the original speed but in the opposite direction.

P10.13. Prepare: Energy is stored in the flywheel by virtue of the motion of the particles and is given by Equation 10.9. In this equation, units for rotational velocity must be rad/s.

Solve: Using Equation 10.92,

$$K_{\text{rot}} = \frac{1}{2}I\omega^2, \text{ so } I = \frac{2K_{\text{rot}}}{\omega^2}$$

We need to convert ω to proper units, radians/s. Since $\omega = 20,000$ rev/min and there are 2π rad/rev and 60 s/min,

$$\omega = \left(20,000\frac{\text{rev}}{\text{min}}\right)\left(\frac{1\text{ min}}{60\text{ s}}\right)\left(\frac{2\pi\text{ rad}}{\text{rev}}\right)$$

So.

$$I = \frac{(2)(4.0\times10^6\text{ J})}{\left[\left(20,000\frac{\text{rev}}{\text{min}}\right)\left(\frac{1\text{ min}}{60\text{ s}}\right)\left(\frac{2\pi\text{ rad}}{\text{rev}}\right)\right]^2} = 1.8\text{ kg}\cdot\text{m}^2$$

Assess: The flywheel can store this large amount of energy even though it has a low moment of inertia because of its high rate of rotation.

P10.15. Prepare: In part **(a)** we can simply use the definition of kinetic energy in Equation 10.8. We then use this result in part **(b)** to find the height the car must be dropped from to obtain the same kinetic energy. The car is falling under the influence of gravity. We can use conservation of energy to calculate its kinetic energy as a result of the fall. The sum of kinetic and potential energy does not change as the car falls.
Solve: **(a)** The kinetic energy of the car is

$$K_{\text{C}} = \frac{1}{2}m_{\text{C}}v_{\text{C}}^2 = \frac{1}{2}(1500\text{ kg})(30\text{ m/s})^2 = 6.8\times10^5\text{ J}$$

We keep one additional significant figure here for use in part **(b)**.
(b) Refer to the diagram.

Here we set K_{f} equal to K_{C} in part **(a)** and place our coordinate system on the ground at $y_{\text{f}} = 0$ m. At this point, the car's potential energy $\left(U_{\text{g}}\right)_{\text{f}}$ is zero, its velocity is v_{f}, and its kinetic energy is K_{f}. At position y_{i}, $v_{\text{i}} = 0$ m/s, so $K_{\text{i}} = 0$ J, and the only energy the car has is $\left(U_{\text{g}}\right)_{\text{i}} = mgy_{\text{i}}$. Since the sum $K + U_{\text{g}}$ is unchanged by motion, $K_{\text{f}} + \left(U_{\text{g}}\right)_{\text{f}} = K_{\text{i}} + \left(U_{\text{g}}\right)_{\text{i}}$. This means

$$K_{\text{f}} + mgy_{\text{f}} = K_{\text{i}} + mgy_{\text{i}} \Rightarrow K_{\text{f}} + 0 = K_{\text{i}} + mgy_{\text{i}}$$

$$\Rightarrow y_{\text{i}} = \frac{(K_{\text{f}} - K_{\text{i}})}{mg} = \frac{(6.75\times10^5\text{ J} - 0\text{ J})}{(1500\text{ kg})(9.80\text{ m/s}^2)} = 46\text{ m}$$

To check if this result depends on the car's mass, rewrite the result of part **(b)** leaving m as a variable, and check if it cancels out.

$$y_i = \frac{(K_f - K_i)}{mg} = \frac{\frac{1}{2}mv_f^2 - \frac{1}{2}mv_i^2}{mg} = \frac{\left(v_f^2 - v_i^2\right)}{2g}$$

Since m cancels out, the distance does *not* depend upon the mass.

Assess: A car traveling at 30 m/s is traveling at 108 km/hr or 67 mi/hr. At that speed, it has the same amount of energy as from being dropped 46 m, which is 151 ft, or from the top of an approximately 19 story building!

P10.19. Prepare: Assume an ideal spring that obeys Hooke's law. Equation 10.15 gives the energy stored in a spring. The elastic potential energy of a spring is defined as $U_s = \frac{1}{2}kx^2$, where x is the magnitude of the stretching or compression relative to the unstretched or uncompressed length. $\Delta U_s = 0$ when the spring is at its equilibrium length and $x = 0$.

Solve: We have $U_s = 200$ J and $k = 1000$ N/m. Solving for x:

$$x = \sqrt{2U_s/k} = \sqrt{2(200 \text{ J})/(1000 \text{ N/m})} = 0.63 \text{ m}$$

Assess: In the equation for the elastic potential energy stored in a spring, it is always the distance of the stretching of compression relative to the *unstretched* or *equilibrium* length.

P10.23.   Prepare: Since the gravitational potential energy and the kinetic energy of the car do not change, all the work Mark does on the car goes into thermal energy.

Solve: The thermal energy created in the tires and the road may be determined by:

$$\Delta E_{th} = W_{Mark} = F_{Mark}d\cos 0° = (110 \text{ N})(150 \text{ m})\cos 0° = 470 \text{ J}$$

Assess: All the work Mark does in pushing the car, becomes thermal energy of the tires and road. Since Mark is pushing in the direction the car is moving, the angle between the direction of F and d is $50°$

P10.29.   Prepare: The following figure shows a before-and-after pictorial representation of the rolling car. The car starts at rest from the top of the hill since it slips out of gear. Since we are ignoring friction, energy is conserved. The total energy of the car at the top of the hill is equal to total energy of the car at any other point.

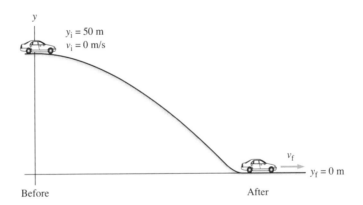

Solve: The energy conservation equation then becomes

$$K_i + (U_g)_i = K_f + (U_g)_f$$

or

$$\frac{1}{2}mv_i^2 + mgy_i = \frac{1}{2}mv_f^2 + mgy_f$$

The car starts from rest, so $v_i = 0$ m/s, which gives $K_i = 0$ J. Taking the bottom of the hill as the reference point for gravitational potential, $y_f = 0$ m and so $U_f = 0$ J.

The energy conservation equation becomes

$$\frac{1}{2}mv_f^2 = mgy_i$$

Canceling m and solving for v_f,

$$v_f = \sqrt{2gy_i} = \sqrt{(2)(9.80 \text{ m/s}^2)(50 \text{ m})} = 31 \text{ m/s}$$

Assess: Note that the problem does not give the shape of the hill, so the acceleration of the car is not necessarily constant. Constant acceleration kinematics can't be used to find the car's final speed. However, energy is conserved no matter what the shape of the hill. Note that the mass of the car is not needed. Since kinetic energy and gravitational potential energy are both proportional to mass, the mass cancels out in the equation. The final speed of the car, after traveling to the bottom of the 50 m hill is 31 m/s which is nearly 70 mi/hr!

P10.31. Prepare: Consider the spring as an ideal spring that obeys Hooke's law. We will also assume zero rolling friction during the compression of the spring, so that mechanical energy is conserved. At the maximum compression of the spring, 60 cm, the velocity of the cart will be zero.

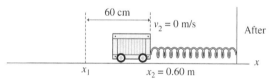

The figure shows a before-and-after pictorial representation. The "before" situation is when the cart hits the spring in its equilibrium position. We put the origin of our coordinate system at this equilibrium position of the free end of the spring. This give $x_1 = x_e = 0$ and $x_2 = 60$ cm.

Solve: The conservation of energy equation $K_2 + U_{s2} = K_1 + U_{s1}$ is

$$\frac{1}{2}mv_2^2 + \frac{1}{2}kx_2^2 = \frac{1}{2}mv_1^2 + \frac{1}{2}kx_1^2$$

Using $v_2 = 0$ m/s, $x_2 = 0.60$ m, and $x_1 = 0$ m gives:

$$\frac{1}{2}kx_2^2 = \frac{1}{2}mv_1^2 \Rightarrow v_1 = \left(\sqrt{\frac{k}{m}}\right)x_2 = \left(\sqrt{\frac{250 \text{ N/m}}{10 \text{ kg}}}\right)(0.60 \text{ m}) = 3.0 \text{ m/s}$$

Assess: Elastic potential energy is always measured from the unstretched or uncompressed length of the spring.

P10.35. Prepare: The thermal energy of the slide and the child's pants changes during the slide. If we consider the system to be the child and slide, total energy is conserved during the slide. The energy transformations during the slide are governed by the conservation of energy equation, Equation 10.4.

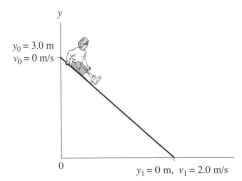

Solve: **(a)** The child's kinetic and gravitational potential energy will be changing during the slide. There is no heat entering or leaving the system, and no external work done on the child. There is a possible change in the thermal energy of the slide and seat of the child's pants. Use the ground as reference for calculating gravitational potential energy.

$$K_i = K_0 = \frac{1}{2}mv_0^2 = 0 \text{ J} \quad U_i = U_{g0} = mgy_0 = (20 \text{ kg})(9.80 \text{ m/s}^2)(3.0 \text{ m}) = 590 \text{ J}$$

$$W = 0 \text{ J} \quad K_f = K_1 = \frac{1}{2}mv_1^2 = \frac{1}{2}(20 \text{ kg})(2.0 \text{ m/s})^2 = 40 \text{ J} \quad U_f = U_{g1} = mgy_1 = 0 \text{ J}$$

At the top of the slide, the child has gravitational potential energy of 590 J. This energy is transformed partly into the kinetic energy of the child at the bottom of the slide. Note that the final kinetic energy of the child is only 40 J, much less than the initial gravitational potential energy of 590 J. The remainder is the change in thermal energy of the child's pants and the slide.

(b) The energy conservation equation becomes $\Delta K + \Delta U_g + \Delta E_{th} = 0.$ With $\Delta U_g = -590$ J and $\Delta K = 40$ J, the change in the thermal energy of the slide and of the child's pants is then 590 J – 40 J = 550 J.

Assess: Note that most of the gravitational potential energy is converted to thermal energy, and only a small amount is available to be converted to kinetic energy.

P10.37. Prepare: This is a one-dimensional collision that obeys the conservation laws of momentum. Since the collision is perfectly elastic, mechanical energy is also conserved. Equation 10.20 applies to perfectly elastic collisions.

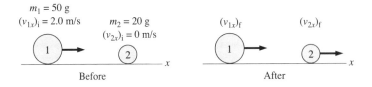

Solve: Using Equation 10.20,

$$\left(v_{1x}\right)_f = \frac{m_1 - m_2}{m_1 + m_2}\left(v_{1x}\right)_i = \frac{50 \text{ g} - 20 \text{ g}}{50 \text{ g} + 20 \text{ g}}(2.0 \text{ m/s}) = 0.86 \text{ m/s}$$

$$\left(v_{1x}\right)_f = \frac{2m_1}{m_1 + m_2}\left(v_{1x}\right)_i = \frac{2(50 \text{ g})}{50 \text{ g} + 20 \text{ g}}(2.0 \text{ m/s}) = 2.9 \text{ m/s}$$

Assess: These velocities are of a reasonable magnitude. Since both these velocities are positive, both balls move along the positive x direction. This makes sense since ball 1 is more massive than ball 2 and ball 2 is initially at rest.

P10.39. Reason: For a perfectly inelastic collision, the two collision objects stick together after the collision and energy is not conserved. Since we are given no information about outside forces acting during the collision, we will assume there are none and that momentum is conserved. Knowing these two pieces of information we can solve the problem.

Solve: Conserving momentum we obtain the velocity of the compound object: $mv = 2mV$ or $V = v/2$
The initial kinetic energy (the kinetic energy of the incident glider) is

$$K_i = mv^2/2$$

The final kinetic energy (the kinetic energy of the combined two gliders) is

$$K_f = (2m)V^2/2 = (2m)(v/2)^2/2 = mv^2/4$$

The final kinetic energy is some fraction f of the initial kinetic energy or $K_f = fK_i$
Solving for the fraction f, obtain

$$f = \frac{K_f}{K_i} = \frac{(mv^2/4)}{(mv^2/2)} = \frac{1}{2}$$

Knowing that the final kinetic energy is one-half the initial kinetic energy, we may conclude that one-half the first glider's kinetic energy is transformed into thermal energy during the collision.
Assess: After a quick first glance at this problem, one might conclude that nothing is given and that the problem can not be solved. After thinking about the concepts involved, the problem can be solved and a numerical value obtained even though no values are given.

P10.41. Prepare: We can use the definition of work, Equation 10.5 to calculate the work you do in pushing the block. The displacement is parallel to the force, so we can use $W = Fd$. Since the block is moving at a steady speed, the force you exert must be exactly equal and opposite to the force of friction.

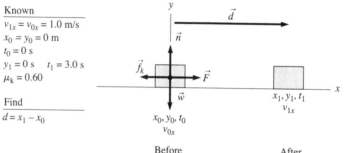

Before   After

Solve: (a) The work done on the block is $W = Fd$ where d is the displacement. We will find the displacement using kinematic equations. The displacement in the *x*-direction is

$$d = (1.0 \text{ m/s})(3.0 \text{ s}) = 3.0 \text{ m}$$

We will find the force using Newton's second law of motion. Consider the preceding diagram.
The equations for Newton's second law along the *x* and *y* components are

$$(F)_y = n - w = 0 \text{ N} \Rightarrow n = w = mg = (10 \text{ kg})(9.80 \text{ m/s}^2) = 98.0 \text{ N}$$

$$(F)_x = \vec{F} - \vec{f_k} = 0 \text{ N} \Rightarrow F = f_k = \mu_k n = (0.60)(98 \text{ N}) = 58.8 \text{ N}$$

$$\Rightarrow W = Fd = (58.8 \text{ N})(3.0 \text{ m}) = 176 \text{ J}, \text{ which should be reported as } 1.8 \times 10^2 \text{ J to two significant figures.}$$

An extra significant figure has been kept in intermediate calculations.
(b) The power required to do this much work in 3.0 s is

$$P = \frac{W}{t} = \frac{176 \text{ J}}{3.0 \text{ s}} = 59 \text{ W}$$

Assess: This seems like a reasonable amount of power to push a 10 kg block at 1.0 m/s. Note that this power is almost what a standard 60 W lightbulb requires!

P10.43. Prepare: The work done on the car while it is accelerating from rest to the final speed is the change in kinetic energy. Knowing the work done and the time to do this work we can determine the power associated with this work.

Solve: The change in kinetic energy of the car is

$$W = \Delta K = K_{\mathrm{f}} - K_{\mathrm{i}} = \frac{1}{2} m v_{\mathrm{f}}^2 = \frac{1}{2}(1000 \text{ kg})(30 \text{ m/s})^2 = 4.5 \times 10^5 \text{ J}$$

since the initial kinetic energy is zero.
The power associated with this work is

$$P = \frac{W}{\Delta t} = \frac{4.5 \times 10^5 \text{ J}}{10 \text{ s}} = 45 \text{ kW}$$

Assess: This is reasonable. In most cars only a small fraction of the work done by the engine goes into propelling the car.

P10.49. Prepare: Assuming that the track offers no rolling friction, the sum of the skateboarder's kinetic and gravitational potential energy does not change during his rolling motion.

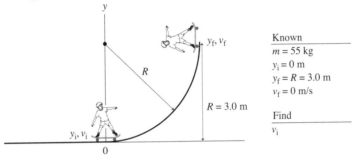

The vertical displacement of the skateboarder is equal to the radius of the track.

Solve: **(a)** The quantity $K + U_{\mathrm{g}}$ is the same at the upper edge of the quarter-pipe track as it was at the bottom. The energy conservation equation $K_{\mathrm{f}} + U_{\mathrm{gf}} = K_{\mathrm{i}} + U_{\mathrm{gi}}$ is

$$\frac{1}{2} m v_{\mathrm{f}}^2 + m g y_{\mathrm{f}} = \frac{1}{2} m v_{\mathrm{i}}^2 + m g y_{\mathrm{i}} \Rightarrow v_{\mathrm{i}}^2 = v_{\mathrm{f}}^2 + 2g(y_{\mathrm{f}} - y_{\mathrm{i}})$$

$$v_{\mathrm{i}}^2 = (0 \text{ m/s})^2 + 2(9.80 \text{ m/s}^2)(3.0 \text{ m} - 0 \text{ m}) = 58.8 \text{ m/s} \Rightarrow v_{\mathrm{i}} = 7.7 \text{ m/s}$$

(b) If the skateboarder is in a low crouch, his height above ground at the beginning of the trip changes to 0.75 m. His height above ground at the top of the pipe remains the same since he is horizontal at that point. Following the same procedure as for part (a),

$$\frac{1}{2} m v_{\mathrm{f}}^2 + m g y_{\mathrm{f}} = \frac{1}{2} m v_{\mathrm{i}}^2 + m g y_{\mathrm{i}} \Rightarrow v_{\mathrm{i}}^2 = v_{\mathrm{f}}^2 + 2g(y_{\mathrm{f}} - y_{\mathrm{i}})$$

$$v_{\mathrm{i}}^2 = (0 \text{ m/s})^2 + 2(9.80 \text{ m/s}^2)(3.0 \text{ m} - 0.75 \text{ m}) = 44.1 \text{ m/s} \Rightarrow v_{\mathrm{i}} = 6.6 \text{ m/s}$$

Assess: Note that we did not need to know the skateboarder's mass, as is the case with free-fall motion. Note that the shape of the track is irrelevant.

P10.53. Prepare: We will need to use Newton's laws here along with the definition of work (Equation 10.5). Assume you lift the box with constant speed.

Solve: **(a)** You lift the box with constant speed so the force you exert must equal the weight of the box. So $F = mg = (20 \text{ kg})(9.80 \text{ m/s}^2) = 196 \text{ N}$. The work done by this force is then

$W = Fd = (196 \text{ N})(1.0 \text{ m}) = 196 \text{ J}$ which should be reported as 0.20 kJ to two significant figures

(b)

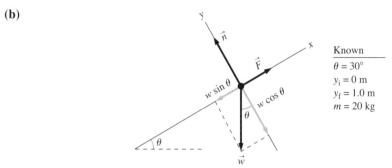

Known
——————
$\theta = 30°$
$y_i = 0$ m
$y_f = 1.0$ m
$m = 20$ kg

Refer to the preceding diagram. Since the box moves at constant speed, the force that is required to push the box up the ramp must exactly equal the component of the gravitational force along the slope.

$$F = mg\sin(\theta) = (20 \text{ kg})(9.80 \text{ m/s}^2)\sin(30°) = 98 \text{ N}$$

(c) Since the height of the ramp is 1.0 m and the angle of the ramp is 30°, the length of the ramp is the length of the hypotenuse in the diagram, which is

$$y = L\sin\theta \Rightarrow L = \frac{y}{\sin\theta} = \frac{1.0 \text{ m}}{\sin 30°} = 2.0 \text{ m}$$

(d) We will use the result of parts (b) and (c) here. The force is parallel to the displacement of the block, so we can use Equation 10.5 again. The work done by the force to push the block up the ramp is $W = Fd = (98 \text{ N})(2.0 \text{ m}) = 196$ J which should be reported as 0.20 kJ to two significant figures.

This is exactly the same result as part (a), where the block is lifted straight up.

Assess: We could have expected that the answers to parts (d) and (a) would be the same. In both cases the force we exert opposes gravity. We know that gravitational potential energy depends only on the change in height of an object, and not the exact path the object follows to change its height. Note that the answer doesn't even depend on the shape of the ramp.

P10.55. Prepare: Since the hill is frictionless, mechanical energy will be conserved during the sledder's trip. To make it over the next hill, the sledder's velocity must be greater than or equal to zero at the top of the hill. The minimum velocity the sledder can have at the top of the second hill is 0 m/s to just make it over. The corresponding velocity at the top of the initial hill will be the minimum the sledder needs to just make it over the next hill.

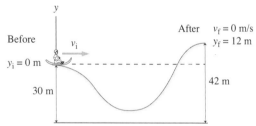

Solve: Consider the before and after pictorial representation. We will use the sledder's initial height as the reference for gravitational potential energy. Since there is no friction, the conservation of energy equation, Equation 10.4 reads

$$\frac{1}{2}mv_i^2 + mgy_i = \frac{1}{2}mv_f^2 + mgy_f \Rightarrow v_i^2 = 2gy_f$$

$$v_i = \sqrt{2gy_f} = \sqrt{2(9.80 \text{ m/s}^2)(12 \text{ m})} = 15 \text{ m/s}$$

Where we have used $y_i = 0$ m, and $v_f = 0$ m/s for the sledder to just make it over the second hill. Note that since we are using the top of the first hill as the reference of gravitational potential energy, we must use the height of the top of the second hill above the first for y_f, $y_f = 42$ m $- 30$ m $= 12$ m.

Assess: Note the shape of the hill doesn't matter, only the difference in height between the first and second hill is needed, as expected for gravitational potential energy. Since the second hill is higher than the first, we expect that the sledder needs the additional kinetic energy at the initial hill to make up for the additional potential energy needed at the top of the second hill.

P10.57.   Prepare: Assume an ideal spring that obeys Hooke's law. There is no friction, and therefore the mechanical energy $K + U_s + U_g$ is conserved. At the top of the slope, as the ice cube is reversing direction, the velocity of the ice cube is 0 m/s.

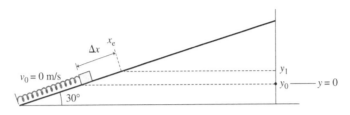

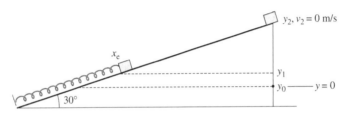

The figure shows a before-and-after pictorial representation. We have chosen to place the origin of the coordinate system at the position where the ice cube has compressed the spring 10.0 cm. That is, $y_0 = 0$.

Solve: The energy conservation equation $K_2 + U_{s2} + U_{g2} = K_0 + U_{s0} + U_{g0}$ is

$$\frac{1}{2}mv_2^2 + \frac{1}{2}k(x_e - x_e)^2 + mgy_2 = \frac{1}{2}mv_0^2 + \frac{1}{2}k(x - x_e)^2 + mgy_0$$

Using $v_2 = 0$ m/s, $y_0 = 0$ m, and $v_0 = 0$ m/s,

$$mgy_2 = \frac{1}{2}k(x - x_e)^2 \Rightarrow y_2 = \frac{k(x - x_e)^2}{2\,mg} = \frac{(25\ \text{N/m})(0.10\ \text{m})^2}{2(0.050\ \text{kg})(9.80\ \text{m/s}^2)} = 26\ \text{cm}$$

The distance traveled along the incline is $y_2 / \sin 30° = 51$ cm.

Assess: The net effect of the launch is to transform the potential energy stored in the spring into gravitational potential energy. The block has kinetic energy as it comes off the spring, but we did not need to know this energy to solve the problem since energy is conserved during the whole process.

P10.63. Reason: If we consider the system of interest to be the two masses and the pulley, the only outside force doing work on the system is gravity. Since gravity is a conservative force, energy is conserved and we may write

$$\Delta K + \Delta U_g = \Delta E = 0 \ \text{ or } \ \Delta K = -\Delta U_g.$$

Solve: Knowing $\Delta K = -\Delta U_g$, we may write

$$(M_A + M_B)v^2 / 2 = -(-M_B gh)$$

or

$$v = \left[2M_B gh / (M_A + M_B)\right]^{1/2} = 1.6\ \text{m/s}$$

Assess: This is a reasonable speed for this situation.

P10.65. Prepare: We can divide this problem into two parts. First, we have an elastic collision between the 20 g ball (m) and the 100 g ball (M). Second, the 100 g ball swings up as a pendulum.

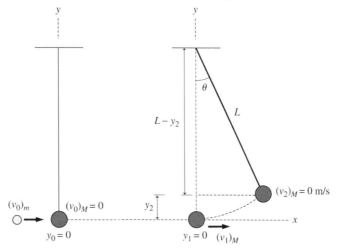

The figure shows three distinct moments of time: the time before the collision, the time after the collision but before the two balls move, and the time the 100 g ball reaches its highest point. We place the origin of our coordinate system on the 100 g ball when it is hanging motionless.

Solve: For a perfectly elastic collision, the ball moves forward with speed

$$(v_1)_M = \frac{2m_m}{m_m + m_M}(v_0)_m = \frac{1}{3.0}(v_0)_m$$

In the second part, the sum of the kinetic and gravitational potential energy is conserved as the 100 g ball swings up after the collision. That is, $K_2 + U_{g2} = K_1 + U_{g1}$. We have

$$\frac{1}{2}M(v_2)_M^2 + Mgy_2 = \frac{1}{2}M(v_1)_M^2 + Mgy_1$$

Using $(v_2)_M = 0$ m/s, $(v_1)_M = \dfrac{(v_0)_m}{3.0}$, $y_1 = 0$ m, and $y_2 = L - L\cos\theta$, the energy equation simplifies to

$$g(L - L\cos\theta) = \frac{1}{2}\frac{(v_0)_m^2}{9.0}$$

$$\Rightarrow (v_0)_m = \sqrt{18\,g\,L(1-\cos\theta)} = \sqrt{18(9.80 \text{ m/s}^2)(1.0 \text{ m})(1-\cos 50°)} = 8.0 \text{ m/s}$$

Assess: Since the collision is elastic, mechanical energy is conserved during the whole process. We could apply conservation of mechanical energy alone to solve this problem. However, solving this particular problem in two parts using momentum conservation for the first part leads to a simpler calculation.

P10.69. Prepare: This is the case of a perfectly inelastic collision. Momentum is conserved because no external force acts on the system (clay block). Mechanical energy is not conserved during perfectly inelastic collisions.

Before   After

$(v_{ix})_1 = v_0$ $(v_{ix})_2 = 0$ m/s v_{fx}

m_1 → m_2 m_1 m_2 →

Solve: (a) The conservation of momentum equation $p_{fx} = p_{ix}$ is

$$(m_1 + m_2)v_{fx} = m_1(v_{ix})_1 + m_2(v_{ix})_2$$

Using $(v_{ix})_1 = v_0$ and $(v_{ix})_2 = 0$, we get

$$v_{fx} = \frac{m_1}{m_1 + m_2}(v_{ix})_1 = \frac{0.050\,\text{kg}}{(1.0\,\text{kg} + 0.050\,\text{kg})}(v_{ix})_1 = 0.048(v_{ix})_1 = 0.048\,v_0$$

(b) The initial and final kinetic energies are given by

$$K_i = \frac{1}{2}m_1(v_{ix})_1^2 + \frac{1}{2}m_2(v_{ix})_2^2 = \frac{1}{2}(0.050\,\text{kg})v_0^2 + \frac{1}{2}(1\,\text{kg})(0\,\text{m/s})^2 = (0.025\,\text{kg})v_0^2$$

$$K_f = \frac{1}{2}(m_1 + m_2)v_{fx}^2 = \frac{1}{2}(1\,\text{kg} + 0.050\,\text{kg})(0.0476)^2 v_0^2 = 0.0012\,v_0^2$$

The percent of energy lost $= \left(\frac{K_i - K_f}{K_i}\right) \times 100\% = \left(1 - \frac{0.0012}{0.025}\right) \times 100\% = 95\%$

The energy goes into the permanent deformation of the ball of clay and into thermal energy.
Assess: Mechanical energy is never conserved during inelastic collisions.

P10.73. Prepare: The motor must pump water to a higher level, and therefore raises the gravitational potential energy of the water. We will calculate the total energy the motor can deliver in one hour, and then use this to calculate the mass of water that can be lifted with this energy.
Solve: Using the conversion 746 W = 1 hp, the motor can put out a power of 1.5 kJ/s. This means $W = Pt = (1.5\,\text{kJ/s})(3600\,\text{s}) = 5.4 \times 10^6\,\text{J}$ is the total work that can be done by the electric motor in one hour. Since all this work goes into giving the water gravitational potential energy,

$$W_{\text{motor}} = U_{gf} - U_{gi} = mg(y_f - y_i) = mg\Delta y$$

$$m = \frac{W_{\text{motor}}}{g\Delta y} = \frac{5.4 \times 10^6\,\text{J}}{(9.80\,\text{m/s}^2)(10\,\text{m})} = 5.5 \times 10^4\,\text{kg}$$

Converting to liters,

$$5.5 \times 10^4\,\text{kg} \times \frac{1\,\text{liter}}{1\,\text{kg}} = 5.5 \times 10^4\,\text{liters}$$

Assess: This seems like a reasonable amount of water for an input power of 2.0 hp.

USING ENERGY

Q11.1. Reason: The friction between your hands increases the kinetic energies in molecules of your hands, and this exhibits itself as an increase in thermal energy of your hands. The temperature of your hands goes up.
Assess: Thermal energy is related to molecular kinetic energies.

Q11.5. Reason: The chicken's efficiency in using the corn's energy is not 100%, so some would be lost if you feed the corn to the chicken. Eat the chicken first, and eat the corn later.
Assess: Your own body isn't 100% efficient in using the corn either, but at least you wouldn't lose twice.

Q11.9. Reason: Since the blocks are at the same temperature, the average kinetic energy of the atoms that make up the block is the same. However, since the 3 kg block contains more atoms, it contains more thermal energy. If the blocks are placed in contact, the average kinetic energy of the atoms in each will remain the same, since there is no source of a higher average kinetic energy. The total energy of the combined system is the sum of the energies of both.
Assess: Thermal energy is directly related to average kinetic energy of the atoms or molecules that make up the system.

Q11.11. Reason: Since the bottles have the same temperature, the average kinetic energies of the atoms that are in each bottle is the same. Since both bottles have the same amount of atoms, the total thermal energy must be the same also. We can also see this from Equation 11.9.
Assess: Thermal energy in a gas is directly proportional to the number of atoms in the gas and its temperature.

Q11.13. Reason: Compressing a gas in an insulated container will increase the thermal energy in the gas without any heat transfer. Work is being done on the gas, so the thermal energy must increase.

Q11.19. Reason: The piston does work on the gas which increases the thermal energy; and it does so quickly enough that heat doesn't escape the system fast enough to keep the gas below ignition temperature.
Assess: Diesel engines have a hard time starting in cold weather for this reason.

Q11.23. Reason: No engine can ever have more efficiency. Any higher efficiency would violate the second law of thermodynamics.

Q11.25. Reason: Yes, the entropy increases as the directed electromagnetic energy is transformed to randomly oriented thermal energy.
Assess: High entropy is typical of random thermal energy.

Problems

P11.1. Prepare: We will use conservation of energy to calculate the energy generated by the engine that is converted entirely to kinetic energy and then use the definition of efficiency to calculate the total energy generated by the engine.
Solve: The conservation of energy states that the work done by the car engine is equal to the change in the kinetic energy. Thus,

$$W_{out} = \Delta K = \tfrac{1}{2}mv^2$$

Using the definition of thermal efficiency,

$$e = \frac{W_{out}}{Q_{H}} \Rightarrow Q_{H} = \frac{W_{out}}{e} = \frac{\frac{1}{2}(1500 \text{ kg})(15 \text{ m/s})^2}{0.10} = 1.7 \times 10^6 \text{ J}$$

That is, the burning of gasoline transfers into the engine 1.7×10^6 J of energy.

Assess: Note that the vast majority of the energy generated by the car's engine is converted to heat. This is typical for engines in cars.

P11.7. Prepare: Various fuels and the corresponding energy in 1 g of each are listed in Table 11.1. Fats in foods such as "energy bars" have an energy content of 38 kJ per gram. Since we have 6 g of fat in our energy bar, we simply need to multiply 6 g by 38 kJ/g.

Solve:

$$6 \text{ g}\left(\frac{38 \text{ kJ}}{1 \text{ g}}\right) = 228 \text{ kJ} \approx 230 \text{ kJ}$$

$$230 \text{ kJ}\left(\frac{1000 \text{ J}}{1 \text{ kJ}}\right) = 230\,000 \text{ J}$$

$$228\,000 \text{ J}\left(\frac{1 \text{ cal}}{4.19 \text{ J}}\right) = 54\,000 \text{ cal}$$

$$54\,000 \text{ cal}\left(\frac{1 \text{ kcal}}{1000 \text{ cal}}\right) = 54 \text{ kcal} = 54 \text{ Cal}$$

Assess: Comparing with Table 11.2 shows that the fat in the energy bar does not provide as much energy as a fried egg; however, the energy bar may also contain carbohydrates in addition, and fat provides more energy per gram than carbohydrates do, so the total number of food calories in the energy bar might be 150.

A 68 kg person needs just over 2000 Cal for basic life processes, so they would need to eat about 15 energy bars per day if that is all they ate.

You may have learned in a health or nutrition class that 1 g of fat provides about 9 Cal of energy. Our calculations above bear this out: $54.4 \text{ Cal}/6 \text{ g} = 9.0 \text{ Cal/g}$.

P11.9. Prepare: Various fuels and the corresponding energy in 1 g of each are listed in Table 11.1. Carbohydrates in foods such as "energy bars" have an energy content of 17 kJ per gram. Since we have 22 g of carbohydrates in our energy bar, we simply need to multiply 22 g by 17 kJ/g.

Table 11.4 tells us that a 68 kg person needs to expend 380 J/s to walk at a speed of 5 km/hr.

Solve: The energy in the carbohydrates in the bar is

$$22 \text{ g}\left(\frac{17 \text{ kJ}}{1 \text{ g}}\right) = 370 \text{ kJ}\left(\frac{1000 \text{ J}}{1 \text{ kJ}}\right) = 370\,000 \text{ J} = 3.7 \times 10^5 \text{ J}$$

The time that the chemical energy will last at the rate of 380 J/s is

$$\Delta t = \frac{\Delta E_{chem}}{P} = \frac{3.7 \times 10^5 \text{ J}}{380 \text{ W}} = 970 \text{ s} = 16 \text{ min} = 0.27 \text{ hr}$$

And the distance that can be covered during this time at 5 km/hr is

$$\Delta x = v\Delta t = (5 \text{ km/hr})(0.27 \text{ hr}) \approx 1.4 \text{ km}$$

Assess: The answer seems to be in the right ball park; we didn't get an answer of just a few cm nor an answer of many km—either of which we would be suspicious of given just one energy bar.

P11.11. Prepare: A typical efficiency for climbing stairs is about 25%, so we can assume that 25% of the chemical energy in the candy bar is transformed to increased potential energy.

$$\Delta U_g = (0.25)(400 \text{ Cal})\left(\frac{1 \text{ kcal}}{1 \text{ Cal}}\right)\left(\frac{1000 \text{ cal}}{1 \text{ kcal}}\right)\left(\frac{4.2 \text{ J}}{1 \text{ cal}}\right) = 4.2 \times 10^5 \text{ J}$$

Solve: Since $\Delta U_g = mg\Delta y$, the height gained is

$$\Delta y = \frac{\Delta_g}{mg} = \frac{4.2 \times 10^5 \text{ J}}{(60 \text{ kg})(9.8 \text{ m/s}^2)} = 710 \text{ m}$$

If we assume that each flight of stairs has a height of 2.7 m (as is done in Example 11.5), this gives

$$\text{Number of flights} = \frac{710 \text{ m}}{2.7 \text{ m}} \approx 260 \text{ flights}$$

Assess: This is more than enough to get to the top of the Empire State Building twice—all fueled by one candy bar! This is a remarkable result.

P11.15. Prepare: Doubling the kinetic energy corresponds to doubling the absolute temperature, so we find the equivalent on the Kelvin scale.
Solve:

$$20\,^\circ\text{C} = 293 \text{ K}$$

Double this to get 586 K and subtract 273 to get back to $313\,^\circ\text{C}$.

Assess: Don't double the Celsius temperature. It seems that $313\,^\circ\text{C}$ is more than double $20\,^\circ\text{C}$ but it works right when converted to the absolute Kelvin scale.

P11.17. Prepare: Solve $\Delta E_{th} = \frac{3}{2}Nk_B\Delta T$ for ΔT.
Solve:

$$\Delta T = \frac{2}{3}\frac{\Delta E_{th}}{Nk_B} = \frac{2}{3}\frac{10 \text{ J}}{(1.0 \times 10^{23})(1.38 \times 10^{-23})} = 4.8 \text{ K} = 4.8\,^\circ\text{C}$$

$$T_f = T_i + \Delta T = 0\,^\circ\text{C} + 4.8\,^\circ\text{C} = 4.8\,^\circ\text{C}$$

Assess: We expected the temperature to rise from the added thermal energy.

P11.19. Prepare: We will use the first law of thermodynamics, Equation 11.12.
Solve: The first law of thermodynamics is

$$\Delta E_{th} = W + Q \Rightarrow -200 \text{ J} = 500 \text{ J} + Q \Rightarrow Q = -700 \text{ J}$$

The negative sign means a transfer of energy from the system to the environment.
Assess: Because $W > 0$ means a transfer of energy into the system, Q must be less than zero and larger in magnitude than W so that $E_{th\,f} < E_{th\,i}$.

P11.23. Prepare: The efficiency of an engine is given by Equation 11.13.
Solve: (a) The work done by the engine per cycle is

$$W_{out} = Q_H - Q_C = 55 \text{ kJ} - 40 \text{ kJ} = 15 \text{ kJ}$$

(b) During each cycle, the heat transferred into the engine is $Q_H = 55$ kJ, and the heat exhausted is $Q_C = 40$ kJ. The thermal efficiency of the heat engine is

$$e = 1 - \frac{Q_C}{Q_H} = 1 - \frac{40 \text{ kJ}}{55 \text{ kJ}} = 0.27$$

Assess: We could have also gotten the answer to part (b) from part (a), $e = W_{out}/Q_H = (15 \text{ J})/(55 \text{ J}) = 0.27$.

P11.25. Prepare: Assume that the heat engine follows a closed cycle. The efficiency of an engine is given by Equation 11.13.
Solve: The engine's efficiency is

$$e = \frac{W_{out}}{Q_H} = \frac{W_{out}}{Q_C + W_{out}} = \frac{200 \text{ J}}{600 \text{ J} + 200 \text{ J}} = 0.25 = 25\%$$

Assess: This is a reasonable efficiency.

P11.29. Prepare: $T_C = 20°C + 273 = 293$ K. The maximum efficiency of a heat engine is $e_{max} = 1 - \dfrac{T_C}{T_H}$.

Solve for T_H.
Solve:

$$T_H = \frac{T_C}{1 - e_{max}} = \frac{293 \text{ K}}{1 - 0.60} = 733 \text{ K} = 460°C$$

Assess: The hot reservoir has to be really quite hot to achieve that efficiency.

P11.31. Prepare: The COP of a refrigerator is given by the equation before Equation 11.15.
Solve: The coefficient of performance of the refrigerator is

$$K = \frac{Q_C}{W_{in}} = \frac{Q_H - W_{in}}{W_{in}} = \frac{50 \text{ J} - 20 \text{ J}}{20 \text{ J}} = 1.5$$

P11.33. Prepare: The COP of a refrigerator is given by the equation just before Equation 11.15.
Solve: **(a)** The heat extracted from the cold reservoir is calculated as follows:

$$COP = \frac{Q_C}{W_{in}} \Rightarrow 4.0 = \frac{Q_C}{50 \text{ J}} \Rightarrow Q_C = 200 \text{ J}$$

(b) The heat exhausted to the hot reservoir is

$$Q_H = Q_C + W_{in} = 200 \text{ J} + 50 \text{ J} = 250 \text{ J}$$

P11.35. Prepare: The efficiency of a Carnot engine (e_{Carnot}) depends only on the temperatures of the hot and cold reservoirs. On the other hand, the thermal efficiency (e) of a heat engine depends on the heats Q_H and Q_C.
Solve: **(a)** According to the first law of thermodynamics, $Q_H = W_{out} + Q_C$. For engine (a), $Q_H = 50$ J, $Q_C = 20$ J and $W_{out} = 30$ J, so the first law of thermodynamics is obeyed. For engine (b), $Q_H = 10$ J, $Q_C = 7$ J and $W_{out} = 4$ J, so the first law is violated. For engine (c) the first law of thermodynamics is obeyed.
(b) For the three heat engines, the maximum or Carnot efficiency is

$$e_{Carnot} = 1 - \frac{T_C}{T_H} = 1 - \frac{300 \text{ K}}{600 \text{ K}} = 0.50$$

Engine (a) has

$$e = 1 - \frac{Q_C}{Q_H} = \frac{W_{out}}{Q_H} = \frac{30 \text{ J}}{50 \text{ J}} = 0.60$$

This is larger than e_{Carnot}, thus violating the second law of thermodynamics. For engine (b),

$$e = \frac{W_{out}}{Q_H} = \frac{4 \text{ J}}{10 \text{ J}} = 0.40 < e_{Carnot}$$

so the second law is obeyed. Engine (c) has a thermal efficiency that is

$$e = \frac{10 \text{ J}}{30 \text{ J}} = 0.33 < e_{Carnot}$$

so the second law of thermodynamics is obeyed.

Assess: The only engine that doesn't violate the first or second law is engine (c).

P11.37. Prepare: We'll show all the possibilities and then directly count the probability that all three balls will be in Box 1.

Solve:

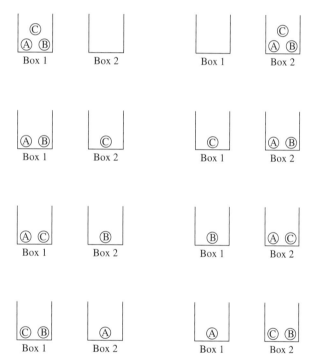

There are eight possible arrangements, and in only one of them are all three balls in Box 1. Therefore, the probability of that happening is $1/8 = 0.125 \approx 13\%$.

Assess: It is not very probable to have all three balls in Box 1 or all in Box 2; much more probable is to have two balls in one box and one ball in the other. This trend becomes more pronounced as the number of balls increases.

P11.39. Prepare: We ignore the energy needed for horizontal motion and simply convert the chemical energy to gravitational potential energy.

Solve:

$$E = \frac{mgh}{e} = \frac{(60 \text{ kg})(9.8 \text{ m/s}^2)(500 \text{ m})}{0.25} = 1.176 \text{ MJ}$$

The total energy divided by the energy per burrito will give the number of burritos.

$$\frac{1.176 \text{ MJ}}{1470 \text{ kJ / burrito}} = 0.80 \text{ burritos}$$

Assess: So a typical hiker could climb a 500 m hill on the energy provided by just less than one frozen burrito.

P11.43. Prepare: Read all the parts of a multipart problem to see the general direction of thought and what preliminary calculations will be needed to get to the final answer.
We convert the time to seconds: 10 minutes and 49 seconds is a total of $649 \text{ s} \approx 650 \text{ s}$.
Solve:
(a) Do a quick calculation for total height: $(86 \text{ floors})(3.7 \text{ m/floor}) = 320 \text{ m} = \Delta y$.

$$\Delta U_g = mg\Delta y = (60 \text{ kg})(9.8 \text{ m/s}^2)(320 \text{ m}) = 187 \text{ kJ} \approx 190 \text{ kJ}$$

This is true for the winner and any other 60 kg person who finished the race.
(b) We'll use the efficiency formula where "what you get" is the change in potential energy of the person, and "what you had to pay" is the total energy the winner expended during the race.

$$\text{what you had to pay} = \frac{\text{what you get}}{25\%} = \frac{187 \text{ kJ}}{0.25} = 748 \text{ kJ} \approx 750 \text{ kJ}$$

(c) This is just unit conversion from the answer in part **(b)**.

$$748 \text{ kJ} \left(\frac{1000 \text{ J}}{1 \text{ kJ}}\right)\left(\frac{1 \text{ cal}}{4.19 \text{ J}}\right)\left(\frac{1 \text{ kcal}}{1000 \text{ cal}}\right) = 179 \text{ kcal} \approx 180 \text{ kcal} = 180 \text{ Cal}$$

(d) Since the winner was 25% efficient, and as we have already noted, 25% of the energy went into increasing the potential energy of the racer. The rest (75%) went into thermal energy.

$$(179 \text{ Cal})(0.75) = 134 \text{ Cal} \approx 130 \text{ Cal}$$

(e) To compute the winner's metabolic power we need to divide the total metabolic energy "burned up" by the time it took.

$$P = \frac{\Delta E}{\Delta t} = \frac{748 \text{ kJ}}{650 \text{ s}} \approx 1200 \text{ W}$$

Assess: The winner of this race obviously made a good effort. In fact, the result of 1200 W is about one and a half horsepower.

P11.47.   Prepare: What we want to get is the 75 W of useful power.

Solve: (a) $\text{what you had to pay} = \dfrac{\text{what you get}}{e} = \dfrac{75 \text{ W}}{0.25} = 300 \text{ W}$

(b) $E = P\Delta t = (300 \text{ W})(1 \text{ h})\left(\dfrac{3600 \text{ s}}{1 \text{ h}}\right) = 1\,080\,000 \text{ J} \approx 1.1 \text{ MJ}$

$$(1\,080\,000 \text{ J})\left(\frac{1 \text{ Cal}}{4190 \text{ J}}\right) = 258 \text{ Cal} \approx 260 \text{ Cal}$$

Assess: That's as much energy as you'd get from a latte with whole milk.

P11.51.   Prepare: The conversion factor is $746 \text{ W} = 1 \text{ hp}$.
Solve:
(a) Because of the 25% efficiency, the horse will need to eat $4 \times 1 \text{ hp}$ worth of Calories.

$$E = P\Delta t = (4 \times 1 \text{ hp})(1 \text{ h})\left(\frac{746 \text{ J}}{1 \text{ hp}}\right)\left(\frac{1 \text{ cal}}{4.19 \text{ J}}\right)\left(\frac{1 \text{ Cal}}{1000 \text{ cal}}\right)\left(\frac{3600 \text{ s}}{1 \text{ h}}\right) = 2564 \text{ Cal} \approx 2600 \text{ Cal}$$

(b) The mass needed would be the total energy divided by the energy per kilogram of hay.

$$m = \frac{2563 \text{ Cal}\left(\dfrac{1000 \text{ cal}}{1 \text{ Cal}}\right)\left(\dfrac{4.19 \text{ J}}{1 \text{ cal}}\right)\left(\dfrac{1 \text{ MJ}}{1 \times 10^6 \text{ J}}\right)}{10 \text{ MJ/kg}} = 1.07 \text{ kg} \approx 1.1 \text{ kg}$$

Assess: A kilogram of food would be a lot for a human, but not so much for a horse.

P11.53. Prepare: We'll use the equation that relates changes in energy with changes in temperature, Equation 11.10, which, fortunately, has N in it, which is what we want to know.
We are not told explicitly that ΔT is negative, but we can conclude so since thermal energy was removed.
Equation 11.10:

$$\Delta E_{\text{th}} = \frac{3}{2} N k_B \Delta T$$

Solve: Solve Equation 11.10 for N.

$$N = \frac{2 \Delta E_{\text{th}}}{3 k_B \Delta T} = \frac{2(-30 \text{ J})}{3(1.38 \times 10^{-23} \text{ J/K})(-20 \text{ K})} = 7.246 \times 10^{22} \text{ atoms} \approx 7.2 \times 10^{22} \text{ atoms}$$

Assess: This is a large number, but typical in problems like this. Avogadro's number is 6.02×10^{23}, so our result is less than the number of carbon atoms in 12 g of carbon.

P11.55. Prepare: The efficiency of an ideal engine (or Carnot engine) depends only on the temperatures of the hot and cold reservoirs.
Solve: (a) The engine's thermal efficiency is

$$e = \frac{W_{\text{out}}}{Q_H} = \frac{W_{\text{out}}}{Q_C + W_{\text{out}}} = \frac{10 \text{ J}}{15 \text{ J} + 10 \text{ J}} = 0.40 = 40\%$$

(b) The efficiency of a Carnot engine is $e_{\text{Carnot}} = 1 - T_C/T_H$. The minimum temperature in the hot reservoir is found as follows:

$$0.40 = 1 - \frac{293 \text{ K}}{T_H} \Rightarrow T_H = 488 \text{ K} = 215°C$$

This is the minimum possible temperature, which should be reported to two significant figures as 220°C.
Assess: In a real engine, the hot-reservoir temperature would be higher than 215°C because no real engine can match the Carnot efficiency.

P11.61. Prepare: Given the efficiency of the power plant we can calculate the amount of waste heat.
Solve: The amount of heat discharged per second is calculated as follows:

$$e = \frac{W_{\text{out}}}{Q_H} = \frac{W_{\text{out}}}{Q_C + W_{\text{out}}} \Rightarrow Q_C = W_{\text{out}}\left(\frac{1}{e} - 1\right) = (900 \text{ MW})\left(\frac{1}{0.32} - 1\right) = 1.9 \times 10^9 \text{ W}$$

That is, each second the electric power plant discharges 1.9×10^9 J of energy into the ocean. Since a typical American house needs 2.0×10^4 J of energy per second for heating, the number of houses that could be heated with the waste heat is $(1.9 \times 10^9 \text{ J})/(2.0 \times 10^4 \text{ J}) = 95,000$.
Assess: This is a relatively efficient electric power plant. Even so, over two times more energy is lost as waste heat than is delivered.

P11.67. Prepare: The maximum possible efficiency of a heat engine is given by Equation 11.14:

$$e_{max} = 1 - \frac{T_C}{T_H}$$

In this problem $T_H = 27°C = 300$ K and $T_C = 3°C = 276$ K

Solve:

$$e_{max} = 1 - \frac{T_C}{T_H} = 1 - \frac{276 \text{ K}}{300 \text{ K}} = 0.080 = 8.0\%$$

Assess: An efficiency of 8% does not sound wonderful, and 1200 m is reasonably deep. But the hot and cold reservoirs are huge, and since nature will maintain the temperatures of the reservoirs for us, if the construction of the heat engine is safe, and cheap, it might be worth it.

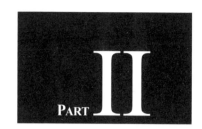
PptII.21. Reason:

$$P = \frac{\Delta E}{\Delta t} = \frac{\frac{1}{2}mv^2}{\Delta t} = \frac{\frac{1}{2}(80 \text{ kg})(11 \text{ m/s})^2}{4.1 \text{ s}} = 1180 \text{ W} \approx 1200 \text{ W}$$

The answer is D.
Assess:

PptII.25. Reason:
(a) Use conservation of momentum.

$$m_L(v_L)_i = (m_L + m_s)v_f \quad \Rightarrow \quad v_f = \frac{m_L(v_L)_i}{m_L + m_s} = \frac{(30 \text{ kg})(4.0 \text{ m/s})}{40 \text{ kg}} = 3.0 \text{ m/s}$$

(b)

$$F = \frac{\Delta p}{\Delta t} = \frac{(30 \text{ kg})(4.0 \text{ m/s})}{0.25 \text{ s}} = 480 \text{ N}$$

(c) Use $K = \frac{1}{2}mv^2$.

$$K_i = \frac{1}{2}(30 \text{ kg})(4.0 \text{ m/s})^2 = 240 \text{J} \qquad K_f = \frac{1}{2}(40 \text{ kg})(3.0 \text{ m/s})^2 = 180 \text{ J}$$

$$\Delta K = K_f - K_i = 180 \text{ K} - 240 \text{ J} = -60 \text{ J}$$

This energy was dissipated as thermal energy.
Assess: In part **(c)** it seems reasonable to "lose" a quarter of the energy.

Q12.1. Reason: The mass of a mole of a substance in grams equals the atomic or molecular mass of the substance. Since neon has an atomic mass of 20, a mole of neon has a mass of 20 g. Since N_2 has a molecular mass of 28, a mole of N_2 has a mass of 28 g. Thus a mole of N_2 has more mass than a mole of neon.

Assess: Even though nitrogen *atoms* are lighter than neon atoms, nitrogen molecules are more massive, so a mole of nitrogen has more mass than a mole of neon.

Q12.5. Reason: From Equation 12.7, we know that the temperature of a gas is directly proportional to the square of the rms speed of the molecules. Thus, doubling the typical speed of molecules in a gas increases the temperature by a factor of four. We also know, from Equation 12.11, that the pressure of a gas is directly proportional to its temperature, so doubling the typical speed of the molecules must also increase the pressure by a factor of four. Increasing the speed of molecules in a gas increases the amount of force a molecule exerts on the wall of the container and the rate of collisions with the walls.

Assess: It makes sense that the pressure depends on the square of the average velocity of molecules in a gas since the force and collision rate are proportional to the velocity of a molecule.

Q12.7. Reason: **(a)** As defined in the chapter, a mole is 6.02×10^{23} basic particles, regardless of which chemical element we have. So there are equal numbers of particles in a mole of helium gas and a mole of oxygen gas.

(b) A mole of helium gas has a mass of 4 g, while a mole of oxygen gas has a mass of 32 g, so one gram of helium gas has $\frac{1}{4} \times 6.02 \times 10^{23}$ particles, while one gram of oxygen gas has $\frac{1}{32} \times 6.02 \times 10^{23}$ particles. Therefore, the gram of helium gas has more particles than the gram of oxygen gas.

Assess: We note that the basic particles for the helium gas are helium *atoms* while the basic particles for the oxygen gas are diatomic oxygen *molecules*. That is why the O_2 molecule has a mass of 32 u.

Q12.11. Reason: Equation 12.11 applies. The number of molecules in the gas is constant since the container is sealed. Equation 12.11 can be written as $p = Nk_B(T/V)$.

(a) If the volume is doubled and the temperature tripled, the pressure increases by a factor of 3/2.

(b) If the volume is halved and the temperature tripled, the pressure increases by a factor of six.

Assess: This makes sense. Increasing the temperature increases the pressure in a gas as does decreasing the volume of the container.

Q12.15. Reason: From Table 12.3, we see that water has a significantly higher coefficient of thermal expansion than steel—about six times as much. As the water and steel get hotter, the water expands six times more than the steel. Thus the water will overflow out of the radiator.

Assess: This seems reasonable since we expect gases to expand more than liquids and liquids to expand more than solids when the temperature increases.

Q12.17. Reason: You are heating both containers (each with n moles of nitrogen gas) and thereby increase the internal energy of each by Q, but the temperatures do not rise by the same amount. See Equations 12.25 and 12.26.

$$Q = nC_V\Delta T_A = nC_P\Delta T_B = 10 \text{ J}$$

Consulting Table 12.6 shows that $C_P > C_V$; therefore $\Delta T_A > \Delta T_B$ and since $(T_A)_i = (T_B)_i$, then $(T_A)_f > (T_B)_f$.

Assess: Some of the energy in container B is used as work done in changing the volume. In container A the volume did not change, so no work was done and all of the energy went into changing the temperature.

Q12.21. Reason: At higher elevations the air pressure is lower. The amount of leavening agent should be reduced, since bubbles will form more readily in the lower pressure. Water will also vaporize at a lower temperature, so the baking temperature should be decreased. Decreasing the temperature requires an increase in the baking times.
Assess: Most non-physicist cooks will suggest increasing the amount of water in a recipe instead of increasing baking time.

Q12.25. Reason: The trees help prevent the energy from being radiated out into space on a cold clear night; the trees reflect back down some of the infrared radiation and keep the ground under them warmer. In contrast, the open ground radiates its thermal energy into space without the "blanket" of the trees or clouds to keep the energy in.
Assess: Gardeners in northern climes know to cover their plants on clear fall nights to keep the radiation in and keep the plants from freezing. These early first frosts in the fall are even called radiation frosts. They take place on clear nights with calm winds. Another type, called advective freeze, occurs when very cold air moves in (by convection); advective freezes can take place with winds and clouds present, and are much harder to protect plants against.

Problems

P12.1. Prepare: For each element, one mole of atoms has a mass in grams equal to the atomic mass number; for example, since the atomic mass number of carbon is 12 then there is one mole of carbon atoms in 12 grams; likewise, there is one mole of argon atoms in 40 grams of argon. See Table 12.1.
Solve: The only catch is that hydrogen gas is diatomic, so one mole of diatomic hydrogen gas molecules has a mass of 2 g.

For hydrogen: $(10 \text{ g})(1 \text{ mol}/2 \text{ g}) = 5 \text{ mol}$

For carbon: $(100 \text{ g})(1 \text{ mol}/12 \text{ g}) = 8.3 \text{ mol}$

For lead: $(50 \text{ g})(1 \text{ mol}/207 \text{ g}) = 0.24 \text{ mol}$

The answer is that 100 g of carbon has the most moles.
Assess: Notice that we count atoms for the solids, but molecules for the diatomic gas.

P12.5. Prepare: The volume is clearly the product of the three length measurements; the issue is converting the units. First multiply $L \times W \times H$ to get the number of cm^3, then convert to m^3.
Solve:

$$V = (200 \text{ cm})(40 \text{ cm})(3.0 \text{ cm}) = 24{,}000 \text{ cm}^3$$

Now remember that while $1 \text{ m} = 100 \text{ cm}$, $1 \text{ m}^3 \neq 100 \text{ cm}^3$. Instead, $1 \text{ m}^3 = 1{,}000{,}000 \text{ cm}^3$.

$$24{,}000 \text{ cm}^3 = (24{,}000 \text{ cm}^3)\left(\frac{1 \text{ m}^3}{1{,}000{,}000 \text{ cm}^3}\right) = 0.024 \text{ m}^3$$

Assess: The answer is small—not a very big fraction of one cubic meter; however, this is reasonable given the small height. The conversion factor comes from $(1 \text{ m}/100 \text{ cm})^3$.

P12.7. Prepare: The absolute pressure is the gauge pressure plus one atmosphere at sea level. $1 \text{ atm} = 14.7 \text{ psi}$.
Solve:

$$p = p_g + 1 \text{ atm} = 35.0 \text{ psi} + 14.7 \text{ psi} = 49.7 \text{ psi}$$

Assess: The difference between p and p_g is due to the fact that your tire gauge measures pressure *differences*.

P12.11. Prepare: In order to use the ideal gas law (Equation 12.11) we need to know the number of helium atoms in the gas.

$$N = nN_A = (7.5 \text{ mol})(6.02 \times 10^{23} \text{ mol}^{-1}) = 4.52 \times 10^{24}$$

$$V = 15 \text{L} \left(\frac{1 \text{m}^3}{1000 \text{L}} \right) = 0.015 \text{m}^3$$

As a further preliminary calculation add 1 atm to the gauge pressure to give the absolute pressure and convert the pressure to SI units.

$$p = p_g + 1 \text{atm} = 65 \text{psi} + 14.7 \text{psi} = 79.7 \text{psi} \left(\frac{1 \text{atm}}{14.7 \text{psi}} \right) \left(\frac{101.3 \text{kPa}}{1 \text{atm}} \right) = 549 \text{kPa}$$

Solve: (a) Solve Equation 12.11 for T.

$$T = \frac{pV}{Nk_B} = \frac{(549 \text{kPa})(0.015 \text{m}^3)}{(4.52 \times 10^{24})(1.38 \times 10^{-23} \text{J/K})} = 132 \text{ K} = -105° \text{ C}$$

(b) Now use Equation 12.5 for K_{ave}.

$$K_{ave} = \frac{3}{2} k_B T = \frac{3}{2} (1.38 \times 10^{-23} \text{J/K})(132 \text{ K}) = 2.7 \times 10^{-21} \text{ J}$$

Assess: The answer to part **(a)** is a cold temperature, but it needs to be to get that much gas in that volume.

P12.13. Prepare: Equation 12.12 applies. We must convert all quantities to SI units.
Solve: Converting units,

$$T = -120 + 273 = 153 \text{ K}$$

$$V = (2.0 \text{ L}) \left(\frac{10^{-3} \text{m}^3}{1 \text{ L}} \right) = 2.0 \times 10^{-3} \text{m}^3$$

Using Equation 12.12,

$$p = \frac{nRT}{V} = \frac{(3.0 \text{ mol})(8.31 \text{ J/(mol} \cdot \text{K}))(153 \text{ K})}{2.0 \times 10^{-3} \text{m}^3} = 1.9 \times 10^6 \text{ Pa}$$

Assess: Note that this is about twenty times atmospheric pressure.

P12.17. Prepare: The gas is assumed to be ideal and it expands isothermally.
Solve: (a) Isothermal expansion means the temperature stays unchanged. That is $T_2 = T_1$.
(b) The before-and-after relationship of an ideal gas under isothermal conditions is

$$\frac{p_1 V_1}{T_1} = \frac{p_2 V_2}{T_1} \Rightarrow p_2 = p_1 \frac{V_1}{V_2} = p_1 \left(\frac{V_1}{2V_1} \right) = \frac{p_1}{2}$$

Assess: The gas has a lower pressure at the larger volume, as we would expect.

P12.19. Prepare: The isobaric heating means that the pressure of the argon gas stays unchanged. Argon gas in the container is assumed to be an ideal gas. We must first convert the volumes and temperatures to SI units with $V_1 = 50 \text{ cm}^3 = 50 \times 10^{-6} \text{ m}^3$, $T_1 = 20°C = (273 + 20)K = 293$ K, and $T_2 = 300°C = (300 + 273)K = 573$ K.
Solve: (a) The container has only argon inside with $n = 0.10$ mol. This produces a pressure

$$p_1 = \frac{nRT_1}{V_1} = \frac{(0.10 \text{ mol})(8.31 \text{ J/(mol} \cdot \text{K}))(293 \text{ K})}{50 \times 10^{-6} \text{ m}^3} = 4.87 \times 10^6 \text{ Pa} = 4870 \text{ kPa}$$

An ideal gas process has $p_2V_2/T_2 = p_1V_1/T_1$. Isobaric heating to a final temperature $T_2 = 300°C = 573\,K$ has $p_2 = p_1$, so the final volume is

$$V_2 = \frac{p_1}{p_2}\frac{T_2}{T_1}V_1 = 1 \times \frac{573}{293} \times 50\ cm^3 = 97.8\ cm^3$$

(b)

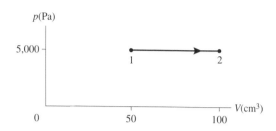

P12.21. Prepare: Assume the gas to be an ideal gas. Please refer to Figure P12.21. We will make use of the following conversions: $1\,atm = 1.013 \times 10^5\,Pa$ and $1\,cm^3 = 1 \times 10^{-6}\,m^3$.
Solve: (a) Because the volume stays unchanged, the process is isochoric.
(b) The ideal-gas law $p_1V_1 = nRT_1$ gives

$$T_1 = \frac{p_1V_1}{nR} = \frac{(3 \times 1.013 \times 10^5\,Pa)(100 \times 10^{-6}\,m^3)}{(0.0040\,mol)(8.31\,J/(mol \cdot K))} = 914\,K$$

The final temperature T_2 is calculated as follows for an isochoric process:

$$\frac{p_1}{T_1} = \frac{p_2}{T_2} \Rightarrow T_2 = T_1\frac{p_2}{p_1} = (914\,K)\left(\frac{1\,atm}{3\,atm}\right) = 300\,K$$

P12.25. Prepare: For a gas in a sealed container (n is constant) we use Equation 12.14.

$$\frac{p_f V_f}{T_f} = \frac{p_i V_i}{T_i}$$

In this case we want to solve for T_f,

$$T_f = \frac{p_f V_f}{p_i V_i}T_i$$

but we get to cancel the volumes since they are the same.
We need to take the usual steps of converting temperatures to the absolute scale and pressures from gauge pressures to absolute pressures.

$$T_i = 0.00°C = 273\,K$$

$$p_i = 55.9\,kPa + 101.3\,kPa = 157.2\,KPa$$

$$p_f = 65.1\,kPa + 101.3\,kPa = 166.4\,KPa$$

Solve:

$$T_f = \frac{p_f}{p_i}T_i = \frac{166.4\,KPa}{157.2\,KPa}273\,K = 289\,K = 16°C$$

Assess: We expected the final temperature to be higher than the initial temperature as the pressure rose. The answer is a reasonable real-life temperature.

P12.29. Prepare: We are given much of the data needed in Equation 12.19 except the coefficient of linear expansion for steel, which we look up in Table 12.3. $\Delta L = 0.73\,mm$, $\Delta T = 13\,K$, $\alpha_{steel} = 12 \times 10^{-6}\,K^{-1}$.

We want to know the original length, so we solve Equation 12.19 for that quantity.
Solve:

$$L_i = \frac{\Delta L}{\alpha \Delta T} = \frac{(0.73 \text{ mm})}{(12 \times 10^{-6} \text{ K}^{-1})(13 \text{ K})}\left(\frac{1 \text{ m}}{1000 \text{ mm}}\right) = 4.7 \text{ m}$$

Assess: This seems to be a reasonable answer—in the realm of daily life, about the width of a room. It would have taken an aluminum beam only about half that long to produce the same ΔL under the same ΔT.

P12.31. Prepare: We need to rearrange Equation 12.2 to give the fractional volume expansion (the percentage expansion is then just 100 times that fractional expansion).
Look up $\beta_{Al} = 69 \times 10^{-6} \text{ K}^{-1}$ in Table 12.3. $\Delta T = 120 \text{ K}$.
Solve:

$$\frac{\Delta V^i}{V_i} = \beta \Delta T = (69 \times 10^{-6} \text{ K}^{-1})(120 \text{ K}) = 0.0083 = 0.83\%$$

Assess: The expansion is small, but greater than it would have been for steel. While 120°C is a larger temperature swing than we might see on a daily basis, there are situations (engines, etc.) where there are significant temperature differences, and people who design and build precise things must take such expansions

P12.33. Prepare: The mass of the mercury is $M = 20 \text{ g} = 2.0 \times 10^{-2} \text{ kg}$, the specific heat from Table 12.4 $c_{mercury} = 140 \text{ J/kg K}$, the boiling point $T_b = 357°C$, and the heat of vaporization from Table 12.5 $L_v = 2.96 \times 10^5 \text{ J/kg}$. We will use Equation 12.22 for obtaining heat needed to raise its temperature to boiling and Equation 12.24 for obtaining heat needed to boil mercury into vapors at the boiling temperature. Note that heating the mercury at its boiling point changes its thermal energy without a change in temperature. We also note that ΔT is the same whether we calculate it in Kelvins or in °C, so we don't have to convert °C into K.
Solve: The heat required for the mercury to change to the vapor phase is the sum of two steps. The first step is

$$Q_1 = Mc_{mercury}\Delta T = (2.0 \times 10^{-2} \text{ kg})(140 \text{ J/(kg} \cdot \text{K)})(357°C - 20°C) = 944 \text{ J}$$

The second step is

$$Q_2 = ML_v = (2.0 \times 10^{-2} \text{ kg})(2.96 \times 10^5 \text{ J/kg}) = 5920 \text{ J}$$

The total heat needed is 6870 J, which will be reported as 6900 J.
Assess: More energy is needed to vaporize mercury (86%) than to warm it to its boiling temperature (14%), as we would expect.

P12.39. Prepare: The water first needs to be raised from 20°C to 100°C (assuming normal atmospheric pressure), and then changed from liquid to gas at the boiling temperature.
We'll use Equation 12.22 $Q = Mc\Delta T$ for the first term and Equation 12.23 $Q = ML$ for the second.
We look up the specific heat of water in Table 12.4 as follows: $c = 4190 \text{ J/(kg} \cdot \text{K)}$. We look up the heat of vaporization of water in Table 12.5: $L_v = 22.6 \times 10^5 \text{ J/kg}$.
We are given $M = 0.10 \text{ kg}$. $\Delta T = 100° - 20° = 80 \text{ K}$.
Solve:

$$Q = Mc\Delta T + ML_v = (0.10 \text{ kg})(4190 \text{ J/(kg} \cdot \text{K)})(80 \text{ K}) + (0.10 \text{ kg})(22.6 \times 10^5 \text{ J/kg})$$
$$= 33,000 \text{ J} + 230,000 \text{ J} = 260,000 \text{ J}$$

Assess: That's quite a few joules, but notice that the vast majority is needed to change the water from liquid to solid rather than raise it from 20°C to 100°C.
Pay attention to the significant figures. The terms that have factors multiplied together each end up with two significant figures, but when those terms are added, the ten-thousand's place is the last significant figure (see Tactics Box 1.1, rule 2).

P12.41. Prepare: A thermal interaction between the copper block and the water leads to a common final temperature denoted by T_f. The initial temperatures of both the copper block and the water are known. The specific heats of copper and water from Table 12.4 are as follows: $c_c = 385 \text{ J/(kg} \cdot \text{K)}$ and $c_w = 4190 \text{ J/(kg} \cdot \text{K)}$.

While the mass of the water is known, we can determine the mass of the copper block using Equations 12.22 and 12.25.

Solve: The conservation of energy equation $Q_{copper} + Q_{water} = 0$ J is

$$M_{copper}c_{copper}(T_f - T_{i\,copper}) + M_{water}c_{water}(T_f - T_{i\,water}) = 0 \text{ J}$$

Both the copper and the water reach the common final temperature $T_f = 25.5°C$. Thus

$$M_{copper}(385 \text{ J/(kg·K)})(25.5°C - 300°C) + 1.0 \text{ kg}(4190 \text{ J/(kg·K)})(25.5°C - 20°C) = 0 \text{ J} \Rightarrow M_{copper} = 0.220 \text{ kg}$$

Assess: Due to the large specific heat of water compared to copper, a smaller value obtained for the mass of the copper block is reasonable.

P12.43. Prepare: We have a thermal interaction between the metal sphere and the mercury that leads to the common final temperature $T_f = 99.0°C$. The initial temperatures of the metal sphere and the mercury are known. The specific heat of mercury from Table 12.4 is $c_{Hg} = 140$ J/(kg·K). The mass of the metal sphere is $M_{metal} = 0.500$ kg and the mass of water is $M_{Hg} = 4.08$ kg. Our strategy is to determine c_{metal} using Equations 12.22 and 12.25.

Solve: The conservation of energy equation $Q_{metal} + Q_{Hg} = 0$ J is

$$M_{metal}c_{metal}(T_f - T_{i\,metal}) + M_{Hg}c_{Hg}(T_f - T_{i\,Hg}) = 0 \text{ J}$$

$$(0.500 \text{ kg})c_{metal}(99°C - 300°C) + (4.08 \text{ kg})(140 \text{ J/(kg·K)})(99°C - 20°C) = 0 \text{ J}$$

We find that $c_{metal} = 449$ J/(kg·K). The metal is iron.

P12.47. Prepare: We will use Equations 12.26 and 12.27 to find the heat needed at constant volume and constant pressure. Since these equations involve the number of moles of the gas, we will calculate it from the mass of the gas and its molar mass. From Table 12.6, $C_P = 20.8$ J/(mol·K) and $C_V = 12.5$ J/(mol·K). Note that the change in temperature on the Kelvin scale is the same as the change in temperature on the Celsius scale.

Solve: **(a)** The atomic mass number of argon is 40. That is, $M_{mol} = 40$ g/mol. The number of moles of argon gas in the container is

$$n = \frac{M}{M_{mol}} = \frac{1.0 \text{ g}}{40 \text{ g/mol}} = 0.025 \text{ mol}$$

The amount of heat is

$$Q = nC_V\Delta T = (0.025 \text{ mol})(12.5 \text{ J/(mol·K)})(100°C) = 31.25 \text{ J}$$

which will be reported as 31 J.

(b) For the isobaric process $Q = nC_P\Delta T$ becomes

$$31.25 \text{ J} = (0.025 \text{ mol})(20.8 \text{ J/(mol·K)})\Delta T \Rightarrow \Delta T = 60°C$$

P12.49. Prepare: From the first law of thermodynamics, $Q = \Delta E_{th} - W$, where W is the work done by the gas. $W = 0$ at constant volume, so, using Equation 12.26, $\Delta E_{th} = Q = nC_V\Delta T$. From Table 12.6, the value of C_V for a monatomic gas is 12.5 J/(mol·K) (which is equal to $3R/2$). For a diatomic gas, we take C_V to be 20.8 J/(mol·K) (which is equal to $5R/2$).

Solve: For a monatomic gas,

$$\Delta E_{th} = nC_V\Delta T = 1.0 \text{ J} = (1.0 \text{ mol})(12.5 \text{ J/(mol·K)})\Delta T \Rightarrow \Delta T = 0.0800°C \text{ or } 0.080 \text{ K}$$

P12.51. Prepare: Please refer to the following figure. The work done by a gas is equal to the area under the pV graph between V_i and V_f. The work done by a gas is positive when $\Delta V > 0$, negative when $\Delta V < 0$, and zero when $\Delta V = 0$. $W_{1\to2} = 0$, $W_{2\to3} =$ Area (I) + Area (II), and $W_{3\to1} = -$Area (II). Thus, the net work done is equal to Area (I). That is, the work done by the gas per cycle is the area inside the *closed* p-versus-V curve. We also need to convert the units of pressure from atm to Pa using the conversion: 1 atm $= 1.013 \times 10^5$ Pa.

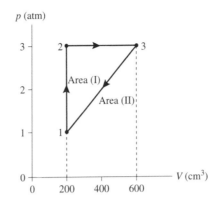

Solve: The area inside the triangle is

$$W_{\text{by gas}} = \frac{1}{2}(3\,\text{atm} - 1\,\text{atm})(600 \times 10^{-6}\,\text{m}^3 - 200 \times 10^{-6}\,\text{m}^3) = \frac{1}{2}\left(2\,\text{atm} \times \frac{1.013 \times 10^5\,\text{Pa}}{1\,\text{atm}}\right)(400 \times 10^{-6}\,\text{m}^3) = 41\,\text{J}.$$

P12.53. Prepare: The rate of conduction across a temperature difference is given in Equation 12.32

$$\frac{Q}{\Delta t} = \left(\frac{kA}{L}\right)\Delta T$$

where $A = 4.0\,\text{m} \times 5.5\,\text{m} = 22\,\text{m}^2$ is the area, $L = 0.018\,\text{m}$ is the thickness of the flooring, and $k = 0.2\,\text{W/(m} \cdot \text{K)}$ is the thermal conductivity of wood given in Table 12.7. $\Delta T = 19.6°\text{C} - 16.2°\text{C} = 3.4°\text{C}$.
Solve:

$$\frac{Q}{\Delta t} = \left(\frac{kA}{L}\right)\Delta T = \frac{(0.2\,\text{W/(m} \cdot \text{K))}(22\,\text{m}^2)}{0.018\,\text{m}}(3.4°\text{C}) = 830\,\text{J/s} = 830\,\text{W}$$

Assess: 830 W is about as much as a dozen incandescent light bulbs. In the winter when you are trying to keep the room warm this energy is being wasted; you could do drastic things like increase the thickness of the wood, or simpler, cheaper things like cover the floor with carpet, which has a much smaller k. In the summer you might be grateful to have this energy conducted from the room if the subfloor can stay at a cooler temperature.

P12.55. Prepare: The rate of energy loss by radiation is given by Equation 12.33.

$$\frac{Q}{\Delta t} = e\sigma A T^4$$

We are given $e = 0.20$, $T = 700°\text{C} = 973$ K, $A = 6 \times (2.0\,\text{cm} \times 2.0\,\text{cm}) = 24\,\text{cm}^2 = 0.0024\,\text{m}^2$. The textbook gives Stefan's constant as $\sigma = 5.67 \times 10^{-8}\,\text{W/(m}^2 \cdot \text{K}^4)$.
Solve:

$$\frac{Q}{\Delta t} = e\sigma A T^4 = (0.20)(5.67 \times 10^{-8}\,\text{W/(m}^2 \cdot \text{K}^4))(0.0024\,\text{m}^2)(973\,\text{K})^4 = 24\,\text{W}$$

Assess: 700°C is quite hot, so the cube radiates a reasonable amount of energy, but even at 700°C the radiation is mostly infrared, not visible. If we double the (absolute) temperature the total radiation would increase by a factor of 16 (due to the T^4) and also a greater portion of the radiation would be in the visible range.

P12.61. Prepare: Treat the gas in the sealed container as an ideal gas and use Equation 12.14.
Solve: **(a)** From the ideal gas law equation $pV = nRT$, the volume V of the container is

$$V = \frac{nRT}{p} = \frac{(2.0\,\text{mol})(8.31\,\text{J/(mol} \cdot \text{K))}[(273 + 30)\,\text{K}]}{1.013 \times 10^5\,\text{Pa}} = 0.050\,\text{m}^3$$

(b) The before-and-after relationship of an ideal gas in a sealed container (constant volume) is

$$\frac{p_1 V}{T_1} = \frac{p_2 V}{T_2} \Rightarrow p_2 = p_1 \frac{T_2}{T_1} = (1.0 \text{ atm}) \frac{(273+130) \text{ K}}{(273+30) \text{ K}} = 1.3 \text{ atm}$$

Assess: Note that gas-law calculations *must* use T in kelvins and pressure *must* be in Pa.

P12.65. Prepare: We find the heat which must be supplied to the air from Equation 12.25 for constant pressure processes. The value of C_p is given in Table 12.6: $C_p = 20.8 \text{ J/(mol·K)}$. The number of moles of air inhaled can be obtained from the ideal gas law. We also find the final volume from the ideal gas law and subtract the initial volume to obtain ΔV.
Solve: (a) Solving the ideal gas law for the number of moles using the initial volume and temperature, we obtain the following:

$$n = \frac{p V_i}{R T_i} = \frac{(1.013 \times 10^5 \text{ Pa})(4.0 \times 10^{-3} \text{ m}^3)}{(8.315 \text{ J/(mol·K)})(273 \text{ K})} = 0.134 \text{ mol}$$

The amount of heat which must be supplied is

$Q = n C_p \Delta T = (0.134 \text{ mol})(29.1 \text{ J/(mol·K)})(37 \text{ K}) = 144 \text{ J} \approx 140 \text{ J.s}$

(b) From the ideal gas law as stated in Equation 12.14, the final volume of the air is given by the following:

$$V_f = \frac{V_i T_f}{T_i} = \frac{(3.0 \text{ L})(310 \text{ K})}{273 \text{ K}} = 3.4 \text{ L}$$

The increase in volume is $\Delta V = V_f - V_i = 0.4 \text{ L}$.

Assess: We could have estimated the change in volume using Equation 12.18, $\Delta V = \beta V_i \Delta T$, but this would not have been as accurate since β depends on temperature (it is given for $T = 20°C$ in Table 12.3) and here the temperature varies from $0°C$ to $37°C$.

P12.67. Prepare: We will use the ideal gas Equation 12.14 and assume that the volume of the tire and that of the air in the tire is constant. That is, the gas undergoes an isochoric (constant-volume) process. Because the gas equation needs absolute rather than gauge pressure, a gauge pressure of 30 psi corresponds to an absolute pressure of (30 psi) + (14.7 psi) = 44.7 psi.
Solve: Using the before-and-after relationship of an ideal gas for an isochoric process,

$$\frac{p_i}{T_i} = \frac{p_f}{T_f} \Rightarrow p_f = \frac{T_f}{T_i} p_i = \left(\frac{273+45}{273+15} \right)(44.7 \text{ psi}) = 49.4 \text{ psi}$$

Your tire gauge will read a gauge pressure $p_f = 49.4 \text{ psi} - 14.7 \text{ psi} = 34.7 \text{ psi}$, which is to be reported as 35 psi.

Assess: A 5 psi increase in gauge pressure due to an increase in temperature by 30°C is reasonable.

P12.69. Prepare: The gas's pressure does not change, so this is an isobaric process. We will use Equation 12.14 with $p_i = p_f$.
Solve: The triple point of water is 0.01°C or 273.16 K, so $T_i = 273.16$ K. Because the pressure is a constant,

$$\frac{V_i}{T_i} = \frac{V_f}{T_f} \Rightarrow T_f = T_i \frac{V_f}{V_i} = (273.16 \text{ K}) \left(\frac{1638 \text{ mL}}{1000 \text{ mL}} \right) = 447.44 \text{ K} = 174°C$$

P12.75. Prepare: Assume CO_2 gas is an ideal gas and Equation 12.12 is applicable. The molar mass for CO_2 is $M_{mol} = 44 \text{ g/mol}$, so a 10 g piece of dry ice is 0.2273 mol. This becomes 0.2273 mol of gas at 0°C.

Solve: **(a)** With $V_1 = 10{,}000\ \text{cm}^3 = 0.010\ \text{m}^3$ and $T_1 = 0°\text{C} = 273\ \text{K}$, the pressure is

$$p_1 = \frac{nRT_1}{V_1} = \frac{(0.2273\ \text{mol})(8.31\ \text{J/(mol}\cdot\text{K)})(273\ \text{K})}{0.010\ \text{m}^3} = 5.156 \times 10^4\ \text{Pa} = 0.509\ \text{atm}$$

or 0.51 atm to two significant figures.

(b) From the isothermal compression, that is, at constant temperature,

$$p_2 V_2 = p_1 V_1 \Rightarrow V_2 = V_1 \frac{p_1}{p_2} = (0.010\ \text{m}^3)\left(\frac{0.509\ \text{atm}}{3.0\ \text{atm}}\right) = 1.70 \times 10^{-3}\ \text{m}^3 = 1700\ \text{cm}^3$$

From the isobaric compression, that is at constant pressure,

$$T_3 = T_2 \frac{V_3}{V_2} = (273\ \text{K})\left(\frac{1000\ \text{cm}^3}{1700\ \text{cm}^3}\right) = 161\ \text{K} = -112°\text{C}$$

or $-110°\text{C}$ to two significant figures.

(c)

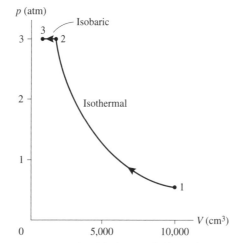

Assess: When volume and pressure appear as a ratio of before and after values, we do not have to convert them to SI units. Temperature, however, must always be converted to Kelvin in the gas equation.

P12.77.   Prepare: The rate of heat transfer of solar energy is $Q/\Delta t$, which is equal to solar power. That is, the heat absorbed is (solar power) $\times \Delta t$ which is also equal to $Mc_{\text{water}}\Delta T$ according to Equation 12.21. From the given information, we can then easily find Δt.

Solve: The area of the garden pond is $A = \pi(2.5\ \text{m})^2 = 19.635\ \text{m}^2$ and the mass of water in the pond is $5.9 \times 10^3\ \text{kg}$. The water absorbs all the solar power, which is

$$(400\ \text{W/m}^2)(19.635\ \text{m}^2) = 7854\ \text{W}$$

This power is used to raise the temperature of the water. That is,

$$Q = (7854\ \text{W})\Delta t = Mc_{\text{water}}\Delta T = (5900\ \text{kg})(4190\ \text{J/(kg}\cdot\text{K)})(10\ \text{K}) \Rightarrow \Delta t = 31{,}476\ \text{s} = 8.7\ \text{h}$$

Assess: It is a common experience that a pool of water takes a few hours to warm up. A value of 8.7 h to heat 5900 kg water by 10°C is reasonable.

P12.81.   Prepare: There are two interacting systems: the nuclear reactor and the water. The heat generated by the nuclear reactor is used to raise the water temperature. For the closed reactor–water system, energy conservation per second requires $Q = Q_{\text{reactor}} + Q_{\text{water}} = 0\ \text{J}$. The heat from the reactor in $\Delta t = 1\ \text{s}$ is $Q_{\text{reactor}} = -2000\ \text{MJ} = -2.0 \times 10^9\ \text{J}$ and we will use Equation 12.21 for Q_{water}.

Solve: The heat absorbed by the water is

$$Q_{water} = m_{water}c_{water}\Delta T = m_{water}(4190 \text{ J/(kg} \cdot \text{K)})(12 \text{ K})$$

$$\Rightarrow -2.0 \times 10^9 \text{ J} + m_{water}(4190 \text{ J/(kg} \cdot \text{K)})(12 \text{ K}) = 0 \text{ J} \Rightarrow m_{water} = 3.98 \times 10^4 \text{ kg}$$

Each second, 3.98×10^4 kg of water is needed to remove heat from the nuclear reactor. Thus, the water flow per minute is

$$3.98 \times 10^4 \frac{\text{kg}}{\text{s}} \times \frac{60 \text{ s}}{\text{min}} = 2.4 \times 10^6 \text{ kg/min}$$

P12.87. Prepare: Heating the material increases its thermal energy. Please refer to Figure P12.87. The material melts at 300°C and undergoes a solid-liquid phase change. The material's temperature increases from 300°C to 1500°C. Boiling occurs at 1500°C and the material undergoes a liquid-gas phase change. We will use Equations 12.21 and 12.21 to determine the specific heat and the heat of vaporization of the liquid.
Solve: **(a)** In the liquid phase, the specific heat of the liquid can be obtained as follows:

$$\Delta Q = Mc\Delta T \Rightarrow c = \frac{1}{M}\frac{\Delta Q}{\Delta T} = \left(\frac{1}{0.200 \text{ kg}}\right)\left(\frac{20 \text{ kJ}}{1200 \text{ K}}\right) = 83 \text{ J/(kg} \cdot \text{K)}$$

(b) The latent heat of vaporization is

$$L_v = \frac{Q}{M} = \frac{40 \text{ kJ}}{(0.200 \text{ kg})} = 2.0 \times 10^5 \text{ J/kg}$$

Assess: The values obtained are of the same order of magnitude as in Tables 12.4 and 12.5 for a few materials.

P12.89. Prepare: There are two interacting systems: aluminum and ice. The system comes to thermal equilibrium in four steps: (1) the ice temperature increases from −10°C to 0°C, (2) the ice becomes water at 0°C, (3) the water temperature increases from 0°C to 20°C, and (4) the cup temperature decreases from 70°C to 20°C. We will use Equations 12.21 and 12.22. The specific heats and the heat of fusion are given in Tables 12.4 and 12.5.
Solve: The aluminum and ice form a closed system, so $Q = Q_1 + Q_2 + Q_3 + Q_4 = 0$ J. These quantities are

$$Q_1 = M_{ice}c_{ice}\Delta T = (0.100 \text{ kg})(2090 \text{ J/(kg} \cdot \text{K)})(10 \text{ K}) = 2090 \text{ J}$$

$$Q_2 = M_{ice}L_f = (0.100 \text{ kg})(3.33 \times 10^5 \text{ J/kg}) = 33,300 \text{ J}$$

$$Q_3 = M_{ice}c_{water}\Delta T = (0.100 \text{ kg})(4190 \text{ J/(kg} \cdot \text{K)})(20 \text{ K}) = 8380 \text{ J}$$

$$Q_4 = M_{Al}c_{Al}\Delta T = M_{Al}(900 \text{ J/(kg} \cdot \text{K)})(-50 \text{ K}) = -(45,000 \text{ J/kg})M_{Al}$$

The $Q = 0$ J equation now becomes

$$43,770 \text{ J} - (45,000 \text{ J/kg})M_{Al} = 0 \text{ J}$$

The solution to this is $M_{Al} = 0.97$ kg.

P12.91. Prepare: There are two interacting systems: coffee (i.e., water) and ice. Changing the coffee temperature from 90°C to 60°C requires four steps: (1) raise the temperature of ice from −20°C to 0°C, (2) change ice at 0°C to water at 0°C, (3) raise the water temperature from 0°C to 60°C, and (4) lower the coffee temperature from 90°C to 60°C. We will use Equations 12.21 and 12.22. The specific heats of water and ice and the heat of fusion of ice are given in Tables 12.4 and 12.5.
Solve: For the closed coffee-ice system,

$$Q = Q_{ice} + Q_{coffee} = (Q_1 + Q_2 + Q_3) + (Q_4) = 0 \text{ J}$$

$$Q_1 = M_{ice}c_{ice}\Delta T = M_{ice}(2090 \text{ J/(kg} \cdot \text{K)})(20 \text{ K}) = M_{ice}(41,800 \text{ J/kg})$$

$$Q_2 = M_{ice}L_f = M_{ice}(330,000 \text{ J/kg})$$

$$Q_3 = M_{ice}c_{water}\Delta T = M_{ice}(4190 \text{ J/kg})(60 \text{ K}) = M_{ice}(251,400 \text{ J/kg})$$

$$Q_4 = M_{coffee}c_{coffee}\Delta T = (300 \times 10^{-6} \text{ m}^3)(1000 \text{ kg/m}^3)(4190 \text{ J/(kg} \cdot \text{K)})(-30 \text{ K}) = -37,000 \text{ J}$$

The $Q = 0$ J equation thus becomes

$$M_{ice}(41,800 + 330,000 + 251,400)\,\text{J/kg} - 37,710\,\text{J} = 0\,\text{J} \Rightarrow M_{ice} = 0.0605\,\text{kg} = 61\,\text{g}$$

Assess: 61 g is the mass of approximately one ice cube.

P12.95. Prepare: The gas is assumed to be an ideal gas that is subjected to isobaric and isochoric processes. Please refer to Figure P12.95. We will use SI units, the ideal gas Equation 12.12, and Equations 12.25 and 12.26.
Solve: (a) The initial conditions are $p_1 = 3.0$ atm $= 304,000$ Pa, $V_1 = 100$ cm$^3 = 1.0 \times 10^{-4}$ m^3, and $T_1 = 100°$C $= 373$ K. The number of moles of gas is

$$n = \frac{p_1 V_1}{R T_1} = \frac{(304,000\,\text{Pa})(1.0 \times 10^{-4}\,\text{m}^3)}{(8.31\,\text{J/(mol} \cdot \text{K})(373\,\text{K})} = 9.81 \times 10^{-3}\,\text{mol}$$

At point 2 we have $p_2 = p_1 = 3.0$ atm and $V_2 = 300$ cm$^3 = 3V_1$. This is an isobaric process, so

$$\frac{V_2}{T_2} = \frac{V_1}{T_1} \Rightarrow T_2 = \frac{V_2}{V_1}T_1 = 3(373\,\text{K}) = 1119\,\text{K}$$

The gas is heated to raise the temperature from T_1 to T_2. The amount of heat required is

$$Q = nC_p\Delta T = (9.81 \times 10^{-3}\,\text{mol})(20.8\,\text{J/mol} \cdot \text{K})(1119\,\text{K} - 373\,\text{K}) = 152\,\text{J}$$

which is to be reported as 150 J.
This amount of heat is *added* during process $1 \rightarrow 2$.
(b) Point 3 returns to $T_3 = 100°$C $= 373$ K. This is an isochoric process, so

$$Q = nC_V\Delta T = (9.81 \times 10^{-3}\,\text{mol})(12.5\,\text{J/mol} \cdot \text{K})(373\,\text{K} - 1119\,\text{K}) = -91.5\,\text{J}$$

which is to be reported as –92 J.
This amount of heat is *removed* during process $2 \rightarrow 3$.

P12.99. Prepare: We'll use Equation 12.33 in a ratio for the two plates. First, solve the equation for T.

$$T = \left[\left(\frac{Q}{\Delta t}\right)\left(\frac{1}{e\sigma A}\right)\right]^{1/4}$$

We are given the rates of energy transfer for the two plates. Since they are the same, and since they are both copper, their emissivities are the same. $A_2 = 4^2 A_1 = 16 A_1$.
Solve:

$$\frac{T_1}{T_2} = \frac{[(\frac{Q}{\Delta t})(\frac{1}{e\sigma A_1})]^{1/4}}{[(\frac{Q}{\Delta t})(\frac{1}{e\sigma A_2})]^{1/4}} = \left(\frac{A_2}{A_1}\right)^{1/4} = \left(\frac{16 A_1}{A_1}\right)^{1/4} = 16^{1/4} = 2$$

Plate 1 is twice as hot as plate 2 on the absolute temperature scale.
Assess: Plate 2 has 16 times the area of plate 1, but because plate 1 is twice as hot as plate 2, they radiate energy at the same rate because of the T^4 in the formula.

FLUIDS

Q13.1. Reason: Density does not depend on the volume. That is, 1 g of mercury would have the same density as 1000 g of mercury, and 1 g of water would have the same density as 1000 g of water.

Table 13.1 shows the density of mercury to be 13,600 kg/m^3 and that of water to be only 1000 kg/m^3.

The density of 1 g of mercury is 13.6 times as much as the density of 1000 g of water.

Assess: It is important to get used to the idea that density is a ratio of mass to volume, so different samples of the same substance would have the same density.

Q13.3. Reason: If the chunk is heavy (dense) enough to sink in water you would put the chunk into a known volume of water in a graduated cylinder and note the rise; this would give the volume of the chunk. A simple measurement of the chunk's mass on a pan balance would then allow you to use $\rho = m/V$.

Assess: If the chunk floats in water then you would have to find a way to submerse it to find the volume.

Q13.7. Reason: The pressure only depends on the depth from the opening. Since A, B, and C are all at the same depth below the opening at E the pressure is the same for each.

$$A = B = C$$

Assess: While there is a taller column of water over B, you can think that the top of the container at D and F pushes down on the fluid too, so the pressure at A, B, and C is the same.

Q13.9. Reason: The ratio of the absolute pressures is smaller. The absolute pressure is the gauge pressure plus atmospheric pressure. For $p_A > p_B$,

$$\frac{p_A}{p_B} = \frac{(p_g)_A + 1\,\text{atm}}{(p_g)_B + 1\,\text{atm}} < \frac{(p_g)_A}{(p_g)_B}$$

Assess: This is easy to see if you plug in a couple of numbers with $p_A > p_B$, say $p_A = 4\,\text{atm}$, $p_B = 3\,\text{atm}$, $(p_g)_A = 3\,\text{atm}$, and $(p_g)_B = 2\,\text{atm}$. $\frac{4}{3} < \frac{3}{2}$.

Q13.15. Reason: The fraction of the object below the surface of the liquid is the object's density as a fraction of the liquid's density (if we can ignore the density of the air). Since A has the greatest fraction below the surface it is the most dense. The least dense (B) floats with the largest fraction of its volume above the fluid level (the smallest fraction below the liquid level).

$$\rho_A > \rho_C > \rho_B$$

Assess: You've heard that only 10% of an iceberg is visible above the surface of the ocean. That means 90% of the iceberg is below the surface. Therefore the density of ice is 90% the density of seawater. Verify this by looking in Table 13.1 and finding on the Web the density of ice, $\rho_{ice} = 917\,\text{kg/m}^3$.

Q13.19. Reason: The scale will read less than your true weight. The scale reads the normal force needed to support your weight, but in the swimming pool your weight is partly supported by the buoyant force. The normal force required from the scale is reduced.

Assess: One way to approach problems like this is to "take limits" and see what the extreme cases tell you. If you and the scale were on a moveable platform, think what the scale would read if it were raised to the surface. The scale reads the normal force, but its magnitude would be the same as your true weight when you are completely out of the water. On the other hand, let the platform slowly lower until you barely have your nose out of water; the scale would read very little. Indeed, if you did this in the Great Salt Lake or the Dead Sea the buoyant force of the salt water would be enough to support your whole body weight and the scale would read zero if the platform is lowered sufficiently.

Q13.21. Reason: Since the volume of the rigid submerged submarine isn't changing, then the buoyant force isn't changing either, but by forcing water out of the tanks the submarine becomes light enough that the buoyant force is greater than the weight. (The submarine with the tanks full of air weighs less than the submarine with the tanks full of water.) The net upward force makes the submarine accelerate upward.
Assess: Equivalently, think of this question in terms of density. To float, the density of the submarine must become less than the density of the water. The two ways to do this are to increase the volume or to decrease the mass. Since the hull of a submarine is rigid the volume doesn't change, but by expelling the mass of the water the submarine becomes less dense—enough to float.

Q13.23. Reason: The sphere is floating in static equilibrium, so the upward buoyant force exactly equals the sphere's weight, $F_B = w$. But according to Archimedes' principle, F_B is the weight of the displaced liquid. That is, the weight of the missing water in B is exactly matched by the weight of the added ball. Thus the total weights of both containers are equal.
Assess: Because the sphere is half below the water line and half above, its density is half that of water. If the sphere were twice as dense (that is, the same density as water) with the same mass, it would occupy half the volume. A hemisphere of water would weigh as much as the original sphere. The hemisphere of water would just fill in the depression left by removal of the sphere, leaving the water height the same as A.

Q13.27. Reason: The Bernoulli effect says that the pressure is higher at the point where the fluid is moving slower, and lower where the fluid is moving faster. The pressure is clearly lowest at point 3 and highest at point 2, so those are the positions of the fastest and slowest fluid, respectively.

$$v_3 > v_1 > v_2$$

Assess: The equation of continuity would then say that the inner diameter of the pipe is greatest at point 2 and smallest at point 3.

Q13.29. Reason: The pressure is reduced at the chimney due to the movement of the wind above. Thus, the air will flow in the window and out the chimney.
Assess: Prairie dogs ventilate their burrows this way; a small breeze above their mound lowers the pressure there and allows the air in the burrow to move between openings of different types or heights.

Problems

P13.1.   Prepare: In SI Units $1 \text{ L} = 10^{-3} \text{ m}^3$ and $1 \text{ g} = 10^{-3} \text{ kg}$.
Solve: The density of the liquid is

$$\rho = \frac{m}{V} = \frac{0.120 \text{ kg}}{100 \text{ mL}} = \frac{0.120 \text{ kg}}{100 \times 10^{-3} \times 10^{-3} \text{ m}^3} = 1200 \text{ kg/m}^3$$

Assess: The liquid's density is a little more than that of water (1000 kg/m^3) and is a reasonable number.

P13.3.   Prepare: The volume of the sphere will be reduced by a factor of 8 when its radius is halved. We will use the definition of mass density $\rho = m/V$.
Solve: The new density is

$$\rho' = \frac{m}{V/8} = 8\frac{m}{V} = 8\rho = 8(1.4 \text{ kg/m}^3) = 11 \text{ kg/m}^3$$

Assess: If the mass is constant and the volume is reduced by a factor of 8, the density will increases by a factor of 8.

P13.5. Prepare: The densities of gasoline and water are given in Table 13.1.
Solve: (a) The total mass is

$$m_{total} = m_{gasoline} + m_{water} = 0.050 \text{ kg} + 0.050 \text{ kg} = 0.100 \text{ kg}$$

The total volume is

$$V_{total} = V_{gasoline} + V_{water} = \frac{m_{gasoline}}{\rho_{gasoline}} + \frac{m_{water}}{\rho_{water}} = \frac{0.050 \text{ kg}}{680 \text{ kg/m}^3} + \frac{0.050 \text{ kg}}{1000 \text{ kg/m}^3} = 1.235 \times 10^{-4} \text{ m}^3$$

$$\Rightarrow \rho_{avg} = \frac{m_{total}}{V_{total}} = \frac{0.100 \text{ kg}}{1.235 \times 10^{-4} \text{ m}^3} = 810 \text{ kg/m}^3$$

(b) The average density is calculated as follows:

$$m_{total} = m_{gasoline} + m_{water} = \rho_{water} V_{water} + \rho_{gasoline} V_{gasoline}$$

$$\Rightarrow \rho_{avg} = \frac{\rho_{water} V_{water} + \rho_{gasoline} V_{gasoline}}{V_{water} + V_{gasoline}} = \frac{(50 \text{ cm}^3)(1000 \text{ kg/m}^3 + 680 \text{ kg/m}^3)}{100 \text{ cm}^3} = 840 \text{ kg/m}^3$$

Assess: The above average densities are between those of gasoline and water, and are reasonable.

P13.9. Prepare: The density of water is 1000 kg/m^3 and the density of ethyl alcohol is 790 kg/m^3. Because the tank is cubic, its side is $d = (V_{water})^{1/3}$.
Solve: (a) The volume of water that has the same mass as 8.0 m^3 of ethyl alcohol is

$$V_{water} = \frac{m_{water}}{\rho_{water}} = \frac{m_{alcohol}}{\rho_{water}} = \frac{\rho_{alcohol} V_{alcohol}}{\rho_{water}} = \left(\frac{790 \text{ kg/m}^3}{1000 \text{ kg/m}^3} \right)(8.0 \text{ m}^3) = 6.32 \text{ m}^3 = 6.3 \text{ m}^3 \text{ to two significant figures}$$

(b) The pressure at the bottom of the cubic tank is $p = p_0 + \rho_{water} g d$:

$$p = 1.013 \times 10^5 \text{ Pa} + (1000 \text{ kg/m}^3)(9.80 \text{ m/s}^2)(6.32)^{1/3} = 1.194 \times 10^5 \text{ Pa} = 1.2 \times 10^5 \text{ Pa to three significant}$$
figures

Assess: Since water is more dense than alcohol, we expect the volume in part (a) to be less. Notice that the contribution of the water to the pressure at the bottom of the tank is small compared to contribution of the atmosphere.

P13.11. Prepare: We'll use Equation 13.5, which gives the pressure of a fluid as a function of depth. We know that the pressure at a given depth depends on the depth but not the diameter of the container; however, we use $d = 5.0$ cm to compute the area of the bottom because $F = pA$. $A = \pi^2 = \pi (2.5 \text{ cm})^2 = 19.63 \text{ cm}^2$.
Solve:

$$F = (p_0 + \rho g d)A$$

$$pA = [101.3 \text{ kPa} + (1000 \text{ kg/m}^3)(9.8 \text{ m/s}^2)(0.35 \text{ m})](19.63 \text{ cm}^2) = [101300 \text{ Pa} + 3430 \text{ Pa}](19.63 \text{ cm}^2)$$

$$= [104730 \text{ Pa}](19.63 \times 10^{-4} \text{ m}^2) = 210 \text{ N}$$

Assess: The weight of the water added a little bit to the weight of the air; you'd have to go down to a depth of almost 11 m to have the contribution due to the weight of the water equal the contribution due to the weight of the air.

P13.15. Prepare: Please refer to Figure P13.15. We assume that there is a perfect vacuum inside the cylinders with $p = 0$ Pa. We also assume that the atmospheric pressure in the room is 1 atm. The flat end of each cylinder has an area $A = \pi r^2 = \pi (0.30 \text{ m})^2 = 0.283 \text{ m}^2$.
Solve: (a) The force on each end is

$$F_{atm} = p_0 A = (1.013 \times 10^5 \text{ Pa})(0.283 \text{ m}^2) = 2.86 \times 10^4 \text{ N} = 2.9 \times 10^4 \text{ N}$$

(b) The net vertical force on the lower cylinder when it is on the verge of being pulled apart is

$$\Sigma F_y = F_{atm} - w_{players} = 0 \text{ N} \Rightarrow w_{players} = F_{atm} = 2.86 \times 10^4 \text{ N} \Rightarrow \text{number of players} = \frac{2.86 \times 10^4 \text{ N}}{(100 \text{ kg})(9.8 \text{ m/s}^2)} = 29.2$$

That is, 30 players are needed to pull the two cylinders apart.

Assess: This problem does an excellent job of helping you understand the ramifications of the fact that we live at the bottom of a very deep air ocean (i.e. atmospheric pressure).

P13.17. Prepare: Please refer to Figure P13.17. Oil is incompressible and has a density of 900 kg/m³.
Solve: (a) The pressure at point A, which is 0.50 m below the open oil surface, is

$$p_A = p_0 + \rho_{oil}g(1.00 \text{ m} - 0.50 \text{ m}) = 101,300 \text{ Pa} + (900 \text{ kg/m}^3)(9.8 \text{ m/s}^2)(0.50 \text{ m}) = 1.1 \times 10^5 \text{ Pa}$$

(b) The pressure difference between A and B is

$$p_B - p_A = (p_0 + \rho g d_B) - (p_0 + \rho g d_A) = \rho g(d_B - d_A) = (900 \text{ kg/m}^3)(9.8 \text{ m/s}^2)(0.50 \text{ m}) = 4400 \text{ Pa}$$

Pressure depends only on depth, and C is the same depth as B. Thus $p_C - p_A = 4400$ Pa also, even though C isn't directly under A.
Assess: This problem illustrates clearly that the pressure depends only on the depth of the fluid.

P13.19. Prepare: Water and mercury are incompressible and immiscible liquids. The water in the left arm floats on top of the mercury and presses the mercury down from its initial level. Because points 1 and 2 are level with each other *and* the fluid is in static equilibrium, the pressure at these two points must be equal. If the pressures were not equal, the pressure difference would cause the fluid to flow, violating the assumption of static equilibrium.

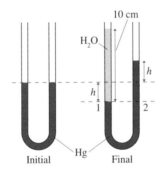

Solve: The pressure at point 1 is due to water of depth $d_w = 10$ cm.

$$p_1 = p_{atmos} + \rho_w g d_w$$

Because mercury is incompressible, the mercury in the left arm goes down a distance h while the mercury in the right arm goes up a distance h. Thus, the pressure at point 2 is due to mercury of depth $d_{Hg} = 2h$.

$$p_2 = p_{atmos} + \rho_{Hg} g d_{Hg} = p_{atmos} + 2\rho_{Hg} g h$$

Equating p_1 and p_2 gives

$$p_{atmos} + \rho_w g d_w = p_{atmos} + 2\rho_{Hg} g h \Rightarrow h = \frac{1}{2}\frac{\rho_w}{\rho_{Hg}}d_w = \frac{1}{2}\frac{1000 \text{ kg/m}^3}{13,600 \text{ kg/m}^3}10 \text{ cm} = 3.68 \text{ mm}$$

Assess: The mercury in the right arm rises 3.68 mm above its initial level. This is a reasonable number due to the rather large density of mercury compared to water.

P13.23. Prepare: We know that the barge displaces an amount of seawater that weighs the same as the barge. This will allow us to compute the weight of the barge. Then, as the barge moves into the fresh water we know that it will also displace an amount of water that weighs the same as the barge. Note that the weight of the displaced seawater, the weight of the displaced fresh water, and the weight of the barge are all the same; their masses are also equal, call that value m. Because fresh water is less dense than seawater (see Table 13.1), the barge will displace a greater volume of fresh water, and the barge will ride lower in the water.

Solve: The volume of seawater displaced is

$$V_{sea} = 3.0 \text{ m} \times 20.0 \text{ m} \times 0.80 \text{ m} = 48 \text{ m}^3$$

The mass of that volume of seawater (and therefore also the mass of the barge) is

$$m = \rho_{sea} V_{sea} = (1030 \text{ kg/m}^3)(48 \text{ m}^3) = 49440 \text{ kg}$$

Now, the volume of fresh water that the barge displaces is

$$V_{fresh} = \frac{m}{\rho_{fresh}} = \frac{49440 \text{ kg}}{1000 \text{ kg/m}^3} = 49.44 \text{ m}^3$$

Lastly, since the area of the barge has not changed, we solve for the new depth.

$$d = \frac{V_{fresh}}{A} = \frac{49.44 \text{ m}^3}{3.0 \text{ m} \times 20.0 \text{ m}} = 0.824 \text{ m} \approx 0.82 \text{ m}$$

In the fresh water the barge rides 2 cm lower than in the seawater.

Assess: The answer is precisely what we expected: The barge rides a bit (2 cm) lower because the fresh water is less dense than the seawater.

In fact, a shortcut would be to see that seawater is 3% more dense than fresh water, so the ship will ride 3% deeper in the fresh water. This gives exactly the same answer: $0.80 \text{ m} \times 103\% = 0.0824 \text{ m}$.

P13.25. Prepare: The buoyant force on the aluminum block is given by Archimedes' principle. The density of aluminum and ethyl alcohol are $\rho_{Al} = 2700 \text{ kg/m}^3$ and $\rho_{ethyl\ alcohal} = 790 \text{ kg/m}^3$. The buoyant force F_B and the tension due to the string act vertically up, and the weight of the aluminum block acts vertically down. The block is submerged, so the volume of displaced fluid equals V_{Al}, the volume of the block.

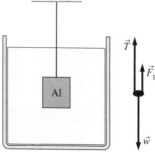

Solve: The aluminum block is in static equilibrium, so

$$\Sigma F_y = F_B + T - w = 0 \text{ N} \Rightarrow \rho_f V_{Al} g + T - \rho_{Al} V_{Al} g = 0 \text{ N} \Rightarrow T = V_{Al} g (\rho_{Al} - \rho_f)$$
$$T = (100 \times 10^{-6} \text{ m}^3)(9.80 \text{ m/s}^2)(2700 \text{ kg/m}^3 - 790 \text{ kg/m}^3) = 1.9 \text{ N}$$

where we have used the conversion $100 \text{ cm}^3 = 100 \times (10^{-2} \text{ m})^3 = 10^{-4} \text{ m}^3$.

Assess: The weight of the aluminum block is $\rho_{Al} V_{Al} g = 2.7 \text{ N}$. A similar order of magnitude for T is reasonable.

P13.29. Prepare: Treat the water as an ideal fluid. Two separate rivers merge to form one river. We will use the definition of flow rate as given by Equation 13.13, $1 \text{ L} = 10^{-3} \text{ m}^3$.

Solve: The volume flow rate in the Bernoulli River is the sum of the volume flow rate of River Pascal and River Archimedes.

$$Q_B = 5.0 \times 10^5 \text{ L/s} + 10.0 \times 10^5 \text{ L/s} = 15.0 \times 10^5 \text{ L/s} = 1500 \text{ m}^3/\text{s}$$

The flowrate is related to the fluid speed and the cross-sectional area by $Q = vA$. Thus

$$v_B = \frac{Q_B}{A_B} = \frac{1500 \text{ m}^3/\text{s}}{150 \text{ m} \times 10 \text{ m}^2} = 1.0 \text{ m/s}$$

Assess: A flow rate of 1.0 m/s is reasonable.

P13.31. Prepare: We refer to Equation 13.11 and write $V = A\Delta x = A(v\Delta t)$. Solve this for Δt. We are given $V = 6000 \text{ L} = 6.0 \text{ m}^3$, $v = 2.1 \text{ m/s}$, and $A = \pi r^2 = \pi\left(\frac{2.5 \text{ cm}}{2}\right)^2 = 4.9 \text{ cm}^2 = 4.9 \times 10^{-4} \text{ m}^2$.

Solve:

$$\Delta t = \frac{V}{Av} = \frac{6.0 \text{ m}^3}{(4.9 \times 10^{-4} \text{ m}^2)(2.1 \text{ m/s})} = 5800 \text{ s} \approx 97 \text{ min}$$

Assess: These all seem like real-life numbers. If you want to empty the pool faster you either need to move the water at a higher speed or get a bigger hose.

P13.33. Prepare: Please refer to Equation 13.14 (Bernoulli's equation). Treat the oil as an ideal fluid obeying Bernoulli's equation. Consider the path connecting point 1 in the lower pipe with point 2 in the upper pipe a streamline.
Solve: Bernoulli's equation is

$$p_2 + \frac{1}{2}\rho v_2^2 + \rho g y_2 = p_1 + \frac{1}{2}\rho v_1^2 + \rho g y_1 \Rightarrow p_2 = p_1 + \frac{1}{2}\rho\left(v_1^2 - v_2^2\right) + \rho g(y_1 - y_2)$$

Using $p_1 = 200 \text{ kPa} = 2.00 \times 10^5 \text{ Pa}$, $\rho = 900 \text{ kg/m}^3$, $y_2 - y_1 = 10.0 \text{ m}$, $v_1 = 2.0 \text{ m/s}$, and $v_2 = 3.0 \text{ m/s}$, we get $p_2 = 1.1 \times 10^5$ Pa = 110 kPa.
Assess: We expect the pressure at point 2 to be less than the pressure at point 1. If this were not the case, the fluid would not flow from point 1 to point 2.

P13.37. Prepare: The equation that relates these variables is Poiseuille's equation. One end of the hose is open to the air, so the gauge pressure requested is the same as Δp.

Known
$Q = 0.25 \text{ L/s} = 2.5 \times 10^{-4} \text{ m}^3/\text{s}$
$R = 1.25 \text{ cm} = 1.25 \times 10^{-2} \text{ m}$
$L = 10 \text{ m}$
$\eta = 1.0 \times 10^{-3} \text{ Pa} \cdot \text{s}$ at 20°C

Find
Δp

Solve: Solve Poiseuille's equation for Δp.

$$\Delta p = \frac{8\eta L Q}{\pi R^4} = \frac{8(1.0 \times 10^{-3} \text{ Pa} \cdot \text{s})(10 \text{ m})(2.5 \times 10^{-4} \text{ m}^3/\text{s})}{\pi(1.25 \times 10^{-2} \text{ m})^4} = 260 \text{ Pa}$$

Assess: This is just a couple of mm of Hg and sounds reasonable. Isn't it nice how the units cancel to leave Pa?

P13.41. Prepare: $N = (M/M_A)N_A$, where N_A is Avogadro's number. Because the atomic mass number of Al is 27, one mole of Al has a mass of $M_A = 27 \text{ g}$.
Solve: The volume of the aluminum cube $V = 8.0 \times 10^{-6} \text{ m}^3$ and its mass is

$$M = \rho V = (2700 \text{ kg/m}^3)(8.0 \times 10^{-6} \text{ m}^3) = 0.0216 \text{ kg} = 21.6 \text{ g}$$

One mole of aluminum (^{27}Al) has a mass of 27 g. The number of atoms is

$$N = \left(\frac{6.02 \times 10^{23} \text{ atoms}}{1 \text{ mol}} \right)\left(\frac{1 \text{ mol}}{27 \text{ g}} \right)(21.6 \text{ g}) = 4.8 \times 10^{23} \text{ atoms}$$

Assess: A number slightly smaller than Avagadro's number is expected since we have slighty less than a mole of aluminum.

P13.43.   Prepare: The pressure at the bottom of the tank is due to the atmosphere, the oil, and the water.
Solve: The pressure at the bottom of the oil layer is

$$p_{\text{bottom}} = p_{\text{o}} + \rho_{\text{oil}}gd_{\text{oil}} + \rho_{\text{water}}gd_{\text{water}}$$

Solving for the height of the oil and inserting values we obtain

$$d_{\text{oil}} = (P_{\text{bottom}} - P_{\text{o}} - \rho_{\text{water}}gd_{\text{water}})/\rho_{\text{oil}}g = 27 cm$$

Assess: Compared to the depth of water, this is a reasonable number.

P13.45. Prepare: Please refer to Figure P13.45. Assume that the oil is incompressible and its density is 900 kg/m^3.
Solve: (a) The pressure at depth d in a fluid is $p = p_0 + \rho gd$. Here, pressure p_0 at the top of the fluid is due both to the atmosphere *and* to the weight of the floating piston. That is, $p_0 = p_{\text{atm}} + w_{\text{p}}/A$. At point A,

$$p_{\text{A}} = P_{\text{atm}} + \frac{w_{\text{p}}}{A} + \rho g(1.00 \text{ m} - 0.30 \text{ m})$$

$$= 1.013 \times 10^5 \text{ Pa} + \frac{(10 \text{ kg})(9.8 \text{ m/s}^2)}{\pi(0.02 \text{ m})^2} + (900 \text{ kg/m}^3)(9.8 \text{ m/s}^2)(0.70 \text{ m}) = 185{,}460 \text{ Pa}$$

$$\Rightarrow F_{\text{A}} = p_{\text{A}}A = (185{,}460 \text{ Pa})\pi(0.10 \text{ m})^2 = 5800 \text{ N}$$

(b) In the same way,

$$p_{\text{B}} = P_{\text{atm}} + \frac{w_{\text{p}}}{A} + \rho g(1.30 \text{ m}) = 190{,}752 \text{ Pa} \Rightarrow F_{\text{B}} = 6000 \text{N}$$

Assess: F_{B} is larger than F_{A}, because p_{B} is larger than p_{A}.

P13.47. Prepare: Assume that the air bubble is always in thermal equilibrium with the surrounding water, and the air in the bubble is an ideal gas.
Solve: The pressure inside the bubble matches the pressure of the surrounding water. At 50 m deep, the pressure is

$$p_1 = p_0 + \rho_{\text{water}}gd = 1.013 \times 10^5 \text{ Pa} + (1000 \text{ kg/m}^3)(9.8 \text{ m/s}^2)(50 \text{ m}) = 5.913 \times 10^5 \text{ Pa}$$

At the lake's surface, $p_2 = p_0 = 1.013 \times 10^5$ Pa. Using the before-and-after relationship of an ideal gas,

$$\frac{p_2 V_2}{T_2} = \frac{p_1 V_1}{T_1} \Rightarrow V_2 = V_1 \frac{p_1}{p_2}\frac{T_2}{T_1} \Rightarrow \frac{4\pi}{3}r_2^3 = \frac{4\pi}{3}(0.005 \text{ m})^3\left(\frac{5.913 \times 10^5 \text{ Pa}}{1.013 \times 10^5 \text{ Pa}} \right)\left(\frac{293 \text{ K}}{283 \text{ K}} \right) \Rightarrow r_2 = 0.0091 \text{ m}$$

The diameter of the bubble is $2r_2 = 0.0182$ m = 1.8 cm.
Assess: The temperature is in degrees kelvin in the ideal gas equation. Because of lower pressure in the bubble at the surface of the lake compared at a depth of 50 m, we expected the bubble to be bigger.

P13.51. Prepare: The buoyant force on the rock is given by Archimedes' principle. We will use Newton's first law, as the rock is in equilibrium. A pictorial representation of the situation and the forces on the rock are shown.

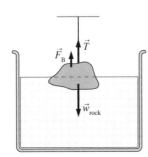

Solve: Because the rock is in static equilibrium, Newton's first law is

$$F_{net} = T + F_B - w_{rock} = 0 \text{ N} \Rightarrow T = \rho_{rock} V_{rock} g - \rho_{water}\left(\frac{1}{2}V_{rock}\right)g =$$

$$\left(\rho_{rock} - \frac{1}{2}\rho_{water}\right)V_{rock}g = \left(\rho_{rock} - \frac{1}{2}\rho_{water}\right)\left(\frac{m_{rock}g}{\rho_{rock}}\right) = \left(1 - \frac{\rho_{water}}{2\rho_{rock}}\right)m_{rock}g$$

Using $\rho_{rock} = 4800$ kg/m^3 and $m_{rock} = 5.0$ kg, we get $T = 44$ N.

Assess: A buoyant force of $(5.0 \times 9.8 \text{ N} - 44 \text{ N}) \approx 5$ N is reasonable for this problem.

P13.55. Prepare: Knowing the volume flow rate, the number of vessels and the diameter of each vessel, we can use Equation 13.13 to solve the problem. We can either say that each vessel handles 1/2000 of the volume flow rate or we can say the total area for this flow rate is 2000 times the area of each vessel. Either way we will get the same result.

Solve: Starting with the definition of volume flow rate and the knowledge that each vessel handles 1/2000 of this volume, we may write

$$\frac{V/2000}{t} = Av$$

Solving for v

$$v = \frac{(V/t)/2000}{\pi r^2} = \frac{3 \times 10^{-8} \text{ m}^3/\text{s}}{(2000)\pi(5 \times 10^{-5} \text{m})^2} = 2 \times 10^{-3} \text{m}/\text{s}$$

Assess: At this rate, water could travel to the top of an 18 m tree in a day. That seems reasonable.

P13.57. Prepare: Please see Figure P13.57. We will treat the water as an ideal fluid obeying Bernoulli's equation. A streamline begins in the bigger size pipe and ends at the exit of the narrower pipe. Let point 1 be beneath the standing column and point 2 be where the water exits the pipe.

Solve: (a) The pressure of the water as it exits into the air is $p_2 = p_{atmos}$.

(b) Bernoulli's equation, Equation 13.14, relates the pressure, water speed, and heights at points 1 and 2.

$$p_1 + \frac{1}{2}\rho v_1^2 + \rho g y_1 = p_2 + \frac{1}{2}\rho v_2^2 + \rho g y_2 \Rightarrow p_1 - p_2 = \frac{1}{2}\rho\left(v_2^2 - v_1^2\right) + \rho g(y_2 - y_1)$$

From the continuity equation,

$$v_1 A_1 = v_2 A_2 = (4\text{m/s})(5 \times 10^{-4} \text{ m}^2) = v_1(10 \times 10^{-4} \text{ m}^2) = 20 \times 10^{-4} \text{ m}^3/\text{s} \Rightarrow v_1 = 2\text{m/s}$$

Substituting into Bernoulli's equation,

$$p_1 - p_2 = p_1 - p_{atmos} = \frac{1}{2}(1000 \text{ kg/m}^3)[(4 \text{ m/s})^2 - (2 \text{ m/s})^2] + (1000 \text{ kg/m}^3)(9.8 \text{ m/s})(4.0 \text{ m})$$

$$= 6000 \text{ Pa} + 39{,}200 \text{ Pa} = 45{,}200 \text{ Pa}$$

But $p_1 - p_2 = \rho g h$, where h is the height of the standing water column. Thus

$$h = \frac{45,200\,\text{Pa}}{(1000\,\text{kg/m}^3)(9.8\,\text{m/s}^2)} = 4.6\,\text{m}$$

Assess: In order to sustain fluid flow, the pressure at point 1 must be greater than the pressure at point 2. As a result we should expect the height h to be greater than 4.0 m.

P13.63. Prepare: The equation that relates these variables is Poiseuille's equation. Solve it for Δp.

$$\Delta p = \frac{8\eta LQ}{\pi R^4}$$

L will be the same in both situations, and we want to keep Q the same as well; this makes us think to use ratios so these will cancel. Call the first situation 1 and the second (restricted arteries) situation 2. $R_2 = 0.90 R_1$. We want to know Δp_2.

Solve:

$$\frac{\Delta p_2}{\Delta p_1} = \frac{\frac{8\eta_2 LQ}{\pi R_2^4}}{\frac{8\eta_1 LQ}{\pi R_1^4}} = \frac{\frac{\eta_2}{R_2^4}}{\frac{\eta_1}{R_1^4}} = \frac{\eta_2}{\eta_1}\left(\frac{R_1}{R_2}\right)^4 = \frac{2.7 \times 10^{-3}\,\text{Pa} \cdot \text{s}}{2.5 \times 10^{-3}\,\text{Pa} \cdot \text{s}}\left(\frac{R_1}{0.90 R_1}\right)^4 = 1.6$$

So $\Delta p_2 = 1.6\Delta p_1 = (1.6)(8.0\,\text{mm Hg}) = 13\,\text{mm Hg}$.

Assess: For the smoker, the required pressure difference is 60% greater than for the non-smoker. This is a significant health issue.

PptIII.19. Reason: Estimate the area of the diaphragm to be $15 \text{ cm} \times 30 \text{ cm} = 0.045 \text{ m}^2$. Since pressure is force/area then

$$F = PA = (7.0 \text{ kPa})(0.045 \text{ m}^2) = 315 \text{ N} \approx 300 \text{ N}$$

Assess: The estimate is probably good to only one significant figure.

PptIII.21. Reason: **(a)** Assume the air in the bladder is an ideal gas. We solve the ideal gas equation for n. The pressure at a depth d is $p = p_0 + \rho g d$. $80 \text{ ft} = 24.384 \text{ m}$. $15°\text{C} = 288 \text{ K}$. $V = (0.070)(7.0 \text{ L}) = 0.00049 \text{ m}^3$.

$$n = \frac{pV}{RT} = \frac{(p_0 + \rho g d)(V)}{RT} = \frac{[101.3 \text{ kPa} + (1030 \text{ kg/m}^3)(9.8 \text{ m/s}^2)(24.384 \text{ m})](0.00049 \text{ m}^3)}{(8.31 \text{ J/(mol·K)})(288 \text{ K})} =$$

$$0.07113 \text{ mol} \approx 0.070 \text{ mol}$$

(b) $50 \text{ ft} = 15.24 \text{ m}$.

$$V = \frac{nRT}{p} = \frac{(71.13 \text{ mol})(8.310 \text{ J/(mol·K)})(288 \text{ K})}{101.3 \text{ kPa} + (1030 \text{ kg/m}^3)(9.8 \text{ m/s}^2)(15.24 \text{ m})} = 0.000667 \text{ m}^3 \approx 6.7 \times 10^{-4} \text{ m}^3$$

(c) Find the new number of moles and then subtract.

$$n = \frac{pV}{RT} = \frac{[101.3 \text{ kPa} + (1030 \text{ kg/m}^3)(9.8 \text{ m/s}^2)(15.24 \text{ m})](0.00049 \text{ m}^3)}{(8.31 \text{ J/(mol·K)})(288 \text{ K})} = 0.05226 \text{ mol}$$

So $0.07113 \text{ mol} - 0.05226 \text{ mol} = 0.01887 \text{ mol} \approx 0.019 \text{ mol}$ need to be removed.

Assess: We expect the needed moles at 50 ft to be less than at 80 ft.

OSCILLATIONS

Q14.1. Reason: There are many examples in daily life, such as a mass hanging from a spring, a tennis ball being volleyed back and forth, washboard road bumps, a beating heart, AC electric voltage, or a pendulum swinging.
Assess: The study of oscillations is important precisely because they occur so often in nature.

Q14.3. Reason: We are given the graph of x versus t. However, we want to think about the slope of this graph to answer velocity questions.
(a) When the x versus t graph is increasing, the particle is moving to the right. It has maximum speed when the positive slope of the x versus t graph is greatest. This occurs at 0 s and 4 s.
(b) When the x versus t graph is decreasing, the particle is moving to the left. It has maximum speed when the negative slope of the x versus t graph is greatest. This occurs at 2 s and 6 s.
(c) The particle is instantaneously at rest when the slope of the x versus t graph is zero. This occurs at 1 s, 3 s, 5 s, and 7 s.
Assess: This is reminiscent of material studied in Chapter 2; what is new is that the motion is oscillatory and the graph periodic.

Q14.5. Reason: The maximum speed of the block is directly proportional to the amplitude. $v_{max} = 2\pi fA$. So doubling the amplitude will double the maximum speed to 40 cm/s.
Assess: We assumed the frequency didn't change.

Q14.9. Reason: From the graph we read the period as 2 s and the maximum speed as 0.10 m/s. Knowing the period we can determine the frequency as follows:

$$f = 1/T = 1/2 \text{ s} = 0.50 \text{ Hz}$$

Knowing the maximum speed and the frequency we can determine the amplitude.

$$v_{max} = \omega A = (2\pi/T)A$$

or

$$A = v_{max}T/2\pi = (0.10 \text{ m/s})(2.0 \text{ s})/2\pi = 0.032 \text{ m}$$

Assess: These are fairly typical values for the frequency and amplitude of an oscillator.

Q14.11. Reason: The period of a simple pendulum is given in Equation 14.31, $T = 2\pi\sqrt{L/g}$. We are told that $T_1 = 2.0$ s.
(a) In this case the mass is doubled, $m_2 = 2m_1$. However, the mass does not appear in the formula for the period of a pendulum; that is, the period does not depend on the mass. Therefore the period is still 2.0 s.
(b) In this case the length is doubled, $L_2 = 2L_1$.

$$\frac{T_2}{T_1} = \frac{2\pi\sqrt{L_2/g}}{2\pi\sqrt{L_1/g}} = \sqrt{\frac{L_2}{L_1}} = \sqrt{\frac{2L_1}{L_1}} = \sqrt{2}$$

So $T_2 = \sqrt{2}T_1 = \sqrt{2}(2.0 \text{ s}) = 2.8$ s.

(c) The formula for the period of a simple small-angle pendulum does not contain the amplitude; that is, the period is independent of the amplitude. Changing (in particular, doubling) the amplitude, as long as it is still small, does not affect the period, so the new period is still 2.0 s.

Assess: It is equally important to understand what *doesn't* appear in a formula. It is quite startling, really, the first time you realize it, that the amplitude (θ_{max}) doesn't affect the period. But this is crucial to the idea of simple harmonic motion. Of course, if the pendulum is swung too far, out of its linear region, then the amplitude would matter. The amplitude *does* appear in the formula for a pendulum not restricted to small angles because the small-angle approximation is not valid; but then the motion is not simple harmonic motion.

Q14.15. Reason: The leg acts somewhat like a pendulum as it swings forward. By bending their knees to bring their feet up closer to the body, sprinters are shortening the pendulums, which makes them swing faster.

Assess: See if you can notice this effect by first running as fast as you can, then again without bending your knees any higher than necessary to clear the ground.

Q14.21. Prepare: When the kangaroo increases its speed the tendons stretch more which stores more energy in them, so they spend more time in the air propelled by greater spring energy.

Assess: Kangaroo hopping can be quite efficient at high speeds.

Problems

P14.1. Prepare: The period of the vibration is the inverse of the frequency, so we will use Equation 14.1.

Solve: The frequency generated by a guitar string is 440 Hz., hence

$$T = \frac{1}{f} = \frac{1}{440\ \text{Hz}} = 2.3 \times 10^{-3}\ \text{s} = 2.3\ \text{ms}$$

Assess: The units of frequency are Hz, that is, cycles per second, or s^{-1}, so the period is in seconds.

P14.5. Prepare: For a small angle pendulum the restoring force is proportional to the displacement because $\sin\theta \approx \theta$.

Solve: If we model this as a Hooke's law situation, then doubling the distance will double the restoring force from 20 N to 40 N.

Assess: This would not work for angles much larger than $10°$.

P14.7. Model: The air-track glider attached to a spring is in simple harmonic motion. The glider is pulled to the right and released from rest at $t = 0$ s. It then oscillates with a period $T = 2.0$ s and a maximum speed $4v_{max} = 0$ cm/s $= 0.40$ m/s. While the amplitude of the oscillation can be obtained from Equation 14.13, the position of the glider can be obtained from Equation 14.10, $x(t) = A\cos\left(\frac{2\pi t}{T}\right)$.

Solve: (a)

$$v_{max} = (2\pi A/T) \Rightarrow A = \frac{v_{max} T}{2\pi} = \frac{(0.40\ \text{m/s})(2.0\ \text{s})}{2\pi} = 0.127\ \text{m} = 0.13\ \text{m}$$

(b) The glider's position at $t = 0.25$ s is

$$x_{0.25\ s} = (0.127\ \text{m})\cos\left[\frac{2\pi(0.25\ \text{s})}{2.0\ \text{s}}\right] = 0.090\ \text{m} = 9.0\ \text{cm}$$

Assess: At $t = 0.25$ s, which is less than one quarter of the time period, the object has not reached the equilibrium position and is still moving toward the left.

P14.9. Prepare: Please refer to Figure P14.9. The oscillation is the result of simple harmonic motion. As the graph shows, the time to complete one cycle (or the period) is $T = 4.0$ s. We will use Equation 14.1 to find frequency.

Solve: (a) The amplitude $A = 20$ cm.

(b) The period $T = 4.0$ s, thus

$$f = \frac{1}{T} = \frac{1}{4.0 \text{ s}} = 0.25 \text{ Hz}$$

Assess: It is important to know how to find information from a graph.

P14.11. Prepare: Treating the building as an oscillator, the magnitude of the maximum displacement is the amplitude, the magnitude of the maximum velocity is determined by $|v_{max}| = 2\pi fA$, and the magnitude of the maximum acceleration is determined by $|a_{max}| = 2\pi^2 f^2 A$.

Solve: The magnitude of the maximum displacement is $|x_{max}| = A = 0.30$ m.

The magnitude of the maximum velocity is $|v_{max}| = 2\pi fA = 2\pi(1.2 \text{ Hz})(0.30 \text{ m}) = 2.3 \text{ m/s}$.

The magnitude of the maximum acceleration is $|a_{max}| = (2\pi f)^2 A = 4\pi^2 (1.2 \text{ Hz})^2 (0.30 \text{ m}) = 17 \text{ m/s}^2$.

Assess: These are reasonable values for the magnitude of the maximum displacement, velocity, and acceleration.

P14.15. Prepare: The spring undergoes simple harmonic motion. The elastic potential energy in a spring stretched by a distance x from its equilibrium position is given by Equation 14.20, and the total mechanical energy of the object is the sum of kinetic and potential energies as in Equation 14.21. At maximum displacement, the total energy is simply $E = \frac{1}{2} kA^2$, Equation 14.22.

Solve: (a) When the displacement is $x = \frac{1}{2} A$, the potential energy is

$$U = \frac{1}{2} kx^2 = \frac{1}{2} k \left(\frac{1}{2} A \right)^2 = \frac{1}{4} \left(\frac{1}{2} kA^2 \right) = \frac{1}{4} E \Rightarrow K = E - U = \frac{3}{4} E$$

Thus, one quarter of the energy is potential and three-quarters is kinetic.

(b) To have $U = \frac{1}{2} E$ requires

$$U = \frac{1}{2} kx^2 = \frac{1}{2} E = \frac{1}{2} \left(\frac{1}{2} kA^2 \right) \Rightarrow x = \frac{A}{\sqrt{2}}$$

P14.17. Prepare: The block attached to the spring is in simple harmonic motion. The period of an oscillating mass on a spring is given by Equation 14.27.

Solve: The period of an object attached to a spring is

$$T = 2\pi \sqrt{\frac{m}{k}} = T_0 = 2.00 \text{ s}$$

where m is the mass and k is the spring constant.

(a) For mass $= 2m$,

$$T = 2\pi \sqrt{\frac{2m}{k}} = (\sqrt{2})T_0 = 2.83 \text{ s}$$

(b) For mass $\frac{1}{2} m$,

$$T = 2\pi \sqrt{\frac{\frac{1}{2} m}{k}} = T_0 / \sqrt{2} = 1.41 \text{ s}$$

(c) The period is independent of amplitude. Thus $T = T_0 = 2.00$ s.

(d) For a spring constant $= 2k$,

$$T = 2\pi\sqrt{\frac{m}{2k}} = T_0/\sqrt{2} = 1.41\ \text{s}$$

Assess: As would have been expected, increase in mass leads to slower simple harmonic motion.

P14.19. Prepare: The oscillating mass is in simple harmonic motion. The position of the oscillating mass is given by $x(t) = (2.0\ \text{cm})\cos(10t)$, where t is in seconds. We will compare this with Equation 14.10.
Solve: (a) The amplitude $A = 2.0$ cm.
(b) The period is calculated as follows:

$$\frac{2\pi}{T} = 10\ \text{rad/s} \Rightarrow T = \frac{2\pi}{10\ \text{rad/s}} = 0.63\ \text{s}$$

(c) The spring constant is calculated from Equation 14.27 as follows:

$$\frac{2\pi}{T} = \sqrt{\frac{k}{m}} \Rightarrow k = m\left(\frac{2\pi}{T}\right)^2 = (0.050\ \text{kg})(10\ \text{rad/s})^2 = 5.0\ \text{N/m}$$

(d) The maximum speed from Equation 14.26 is

$$v_{\text{max}} = 2\pi fA = \left(\frac{2\pi}{T}\right)A = (10\ \text{rad/s})(2.0\ \text{cm}) = 20\ \text{cm/s}$$

(e) The total energy from Equation 14.22 is

$$E = \frac{1}{2}kA^2 = \frac{1}{2}(5.0\ \text{N/m})(0.02\ \text{m})^2 = 1.0\times10^{-3}\ \text{J}$$

(f) At $t = 0.40$ s, the velocity from Equation 14.12 is

$$v_x = -(20.0\ \text{cm/s})\sin[(10\ \text{rad/s})(0.40\ \text{s})] = 15\ \text{cm/s}$$

Assess: Velocity at $t = 0.40$ s is less than the maximum velocity, as would be expected.

P14.23. Prepare: Assume a small angle of oscillation so there is simple harmonic motion. We will use Equation 14.31 for the pendulum's time period.
Solve: The period of the pendulum is

$$T_0 = 2\pi\sqrt{\frac{L_0}{g}} = 4.00\ \text{s}$$

(a) The period is independent of the mass and depends only on the length. Thus $T = T_0 = 4.00$ s.
(b) For a new length $L = 2L_0$,

$$T = 2\pi\sqrt{\frac{2L_0}{g}} = \sqrt{2}T_0 = 5.66\ \text{s}$$

(c) For a new length $L = L_0/2$,

$$T = 2\pi\sqrt{\frac{L_0/2}{g}} = \frac{1}{\sqrt{2}}T_0 = 2.83\ \text{s}$$

(d) The period is independent of the amplitude as long as there is simple harmonic motion. Thus $T = 4.00$ s.

P14.29. Prepare: To complete a whole period, the wrecking ball will have to swing down, up to the other side, back down, and up again to the original position. So the time it takes to swing from maximum height down to lowest height once is one-quarter of a period. We will assume that the wrecking ball is a simple small-angle pendulum and so it's period is given by $T = 2\pi\sqrt{L/g}$.

Solve:

$$\frac{1}{4}T = \frac{1}{4}2\pi\sqrt{\frac{L}{g}} = \frac{\pi}{2}\sqrt{\frac{10\text{ m}}{9.8\text{ m/s}^2}} = 1.6\text{ s}$$

Assess: This *is* enough time to dive out of the way, but it is still wiser to not stand in the way of wrecking balls.

P14.33. Prepare: Treating the hoop suspended from the nail as a physical pendulum we can determine the period of oscillation using $T = 2\pi\sqrt{I/mgL}$.

Solve: The period of oscillation for the hoop suspended from its rim on a nail may be determined by

$$T = 2\pi\sqrt{I/mgL} = 2\pi\sqrt{2mR^2/mgR} = 2\pi\sqrt{2R/g} = 2\pi\sqrt{2(0.22\text{ m})/(9.80\text{ m/s}^2)} = 1.3\text{ s}$$

Assess: This is a reasonable period for this situation.

P14.37. Prepare: The initial amplitude is $A = 6.5$ cm.
Solve:

The equation is $x_{\max}(t) = Ae^{-t/\tau}$ with $x_{\max}(8.0\text{ s}) = 1.8$ cm.

(a)
$$1.8\text{ cm} = (6.5\text{ cm})e^{-8.0\text{ s}/\tau} \quad\Rightarrow\quad \tau = -\frac{8.0\text{ s}}{\ln\frac{1.8\text{ cm}}{6.5\text{ cm}}} = 6.23\text{ s} \approx 6.2\text{ s}$$

(b) Use $x_{\max}(t) = Ae^{-t/\tau}$ with $t = 4.05$ and $\tau = 6.23$ s

$$x_{\max}(4.0\text{ s}) = (6.5\text{ s})e^{-(4.0\text{ s})/(6.23\text{ s})} = 3.4\text{ cm}$$

Assess: We expect the amplitude at 4.0 s to be between 6.5 cm and 1.8 cm.

P14.39. Prepare: We will model the child on the swing as a simple small-angle pendulum. To make the amplitude grow large quickly we want to drive (push) the oscillator (child) at the natural resonance frequency. In other words, we want to wait the natural period between pushes.
Solve:

$$T = 2\pi\sqrt{\frac{L}{g}} = 2\pi\sqrt{\frac{2.0\text{ m}}{9.8\text{ m/s}^2}} = 2.8\text{ s}$$

Assess: You could also increase the amplitude by pushing every other time (every $2T$), but that would not make the amplitude grow as quickly as pushing every period.
The mass of the child was not needed; the answer is independent of the mass.

P14.43. Prepare: Since all forces are conservative, conservation of energy may be used to solve the problem. The following sketch will be helpful in the process of thinking about this problem.

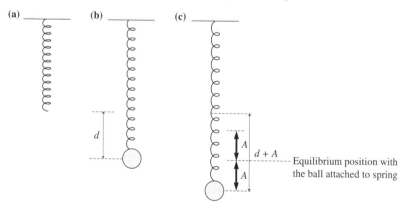

In the figure,

(a) shows the unstretched spring—this is the point of zero potential energy for the spring.

(b) shows the spring with the ball hanging at rest. The spring has been stretched an amount d and it is now at rest at the new equilibrium position. At this position the spring is stretched an amount d until the restoring force the spring exerts is equal to the weight of the ball. That is $kd = mg$ or $d = mg/k$. As a matter of convenience let's choose this new equilibrium position as the point for zero potential energy. Since we are free to choose zero gravitational energy at any point we wish, this seems like a convenient choice.

(c) shows the ball pulled down an additional amount A.

Next let's write a statement of total energy at the stretched position and at the equilibrium position. Since energy is conserved, we will equate these expressions for total energy and solve for the speed of the ball at the equilibrium position, which is the maximum speed (because the total potential energy is zero at this point).

Solve: The initial total energy when the ball is at the stretched position is $E_i = K + U_g + U_s = 0 - mgA + k(d+A)^2/2$.

The final total energy when the ball is at the new spring equilibrium position is $E_f = K + U_g + U_s = mv^2/2 + 0 + kd^2/2$.

Equating the initial and final total energy we have $mv^2/2 + kd^2/2 = -mgA + k(d+A)^2/2$.

Expanding this obtains $mv^2/2 + kd^2/2 = -mgA + k(d^2 + 2dA + A^2)/2$.

Multiplying by 2 and doing a little rearranging, obtain $mv^2 + kd^2 = -2mgA + kd^2 + 2kdA + kA^2$.

Cancel the kd^2 term and insert $d = mg/k$ to obtain $mv^2 = -2mgA + 2k(mg/k)A + kA^2$.

After canceling the $2mgA$ terms and solving for v, obtain

$$v = \sqrt{k/m}\, A = \sqrt{(12 \text{ N/m})/(0.40 \text{ kg})}\,(0.20 \text{ m}) = 1.1 \text{ m/s}$$

Assess: It is interesting to note that we ended up with the same result as the horizontal case.

P14.49. Prepare: The vertical mass/spring systems are in simple harmonic motion. Please refer to Figure P14.49.

Solve: (a) For system A, y is positive for one second as the mass moves downward and reaches maximum negative y after two seconds. It then moves upward and reaches the equilibrium position, $y = 0$, at $t = 3$ seconds. The maximum speed while traveling in the upward direction thus occurs at $t = 3.0$ s. The frequency of oscillation is 0.25 Hz.

(b) For system B, all the mechanical energy is potential energy when the position is at maximum amplitude, which for the first time is at $t = 1.5$ s. The time period of system B is thus 6.0 s.

(c) Spring/mass A undergoes three oscillations in 12 s, giving it a period $T_A = 4.0$ s. Spring/mass B undergoes two oscillations in 12 s, giving it a period $T_B = 6.0$ s. From Equation 14.27, we have

$$T_A = 2\pi\sqrt{\frac{m_A}{k_A}} \quad \text{and} \quad T_B = 2\pi\sqrt{\frac{m_B}{k_B}} \Rightarrow \frac{T_A}{T_B} = \sqrt{\left(\frac{m_A}{m_B}\right)\left(\frac{k_B}{k_A}\right)} = \frac{4.0 \text{ s}}{6.0 \text{ s}} = \frac{2}{3}$$

If $m_A = m_B$, then

$$\frac{k_B}{k_A} = \frac{4}{9} \Rightarrow \frac{k_A}{k_B} = \frac{9}{4} = 2.25$$

Assess: It is important to learn how to read a graph.

P14.51. Prepare: The ball attached to the spring is in simple harmonic motion. The position and velocity at time t are $x_0 = -5$ cm and $v_0 = 20$ cm/s. An examination of Equations 14.19 shows that $x(t) = A\cos(2\pi ft)$ and $\frac{v_x(t)}{2\pi f} = -A\sin(2\pi ft)$. Adding the squares of these equations and using the trigonometric relationship $\cos^2\theta + \sin^2\theta = 1$, we have $A = \sqrt{x(t)^2 + \left(\frac{v_x(t)}{2\pi f}\right)^2}$.

Solve: (a) The oscillation frequency is

$$f = \frac{1}{2\pi}\sqrt{\frac{k}{m}} = \frac{1}{2\pi}\sqrt{\frac{2.5\ \text{N/m}}{0.10\ \text{kg}}} = 0.796\ \text{Hz}$$

Using Equation 14.27, the amplitude of the oscillation is

$$A = \sqrt{x(t)^2 + \left(\frac{v(t)}{2\pi f}\right)^2} = \sqrt{(-5.00\ \text{cm})^2 + \left(\frac{20\ \text{cm/s}}{2\pi(0.796\ \text{Hz})}\right)^2} = 6.40\ \text{cm}$$

(b) We can use the conservation of energy between $x_i = -5$ cm and $x_f = 3$ cm as follows:

$$\frac{1}{2}mv_i^2 + \frac{1}{2}kx_i^2 = \frac{1}{2}mv_f^2 + \frac{1}{2}kx_f^2 \Rightarrow v_f = \sqrt{v_i^2 + \frac{k}{m}\left(x_i^2 - x_f^2\right)} = 0.283\ \text{m/s} = 28.3\ \text{cm/s}$$

Assess: Because k is known in SI units of N/m, the energy calculation *must* be done using SI units of m, m/s, and kg. Both the amplitude and speed are reasonable.

P14.53. Prepare: The compact car is in simple harmonic motion. The mass on each spring for the empty car is $(1200\ \text{kg})/4 = 300$ kg. However, the car carrying four persons means that each spring has, on the average, an additional mass of 70 kg. For both parts we will use Equation 14.27.
Solve: (a) The spring constant can be calculated as follows:

$$f = \frac{1}{2\pi}\sqrt{\frac{k}{m}} \Rightarrow k = m(2\pi f)^2 = (300\ \text{kg})[2\pi(2.0\ \text{Hz})]^2 = 4.74\times10^4\ \text{N/m} \approx 4.7\times10^4\ \text{N/m}$$

to two significant figures.
(b) Here $m = 370$ kg, so

$$f = \frac{1}{2\pi}\sqrt{\frac{k}{m}} = \frac{1}{2\pi}\sqrt{\frac{4.74\times10^4\ \text{N/m}}{370\ \text{kg}}} = 1.8\ \text{Hz}$$

Assess: A small frequency change from the additional mass is reasonable because frequency is inversely proportional to the square root of the mass.

P14.57. Prepare: When the block is displaced from the equilibrium position one spring is compressed and exerts a restoring force on the block while the other spring is stretched and also exerts a restoring force on the block. These two forces have the same magnitude since the springs are identical and a stretch or compression of the same amount produces the same restoring force.
The restoring force would be the same if the springs were on the same side of the block. For springs in parallel like that, the total k_{tot} is the sum of the two k's. The restoring force would even be the same if there were only one spring with a spring constant twice as big (since the two springs each have the same k).
Solve: f_{two} is the frequency for the system with two springs.

$$f_{two} = \frac{1}{2\pi}\sqrt{\frac{k_1 + k_2}{m}} = \frac{1}{2\pi}\sqrt{\frac{20\ \text{N/m} + 20\ \text{N/m}}{2.5\ \text{kg}}} = 0.64\ \text{Hz}$$

Assess: The answer seems reasonable.
If the two k's are the same $k_1 = k_2 = k$ (as in this case), you can see the general formula would be

$$f_{two} = \frac{1}{2\pi}\sqrt{\frac{k_1 + k_2}{m}} = \frac{1}{2\pi}\sqrt{\frac{2k}{m}} = \sqrt{2}\left(\frac{1}{2\pi}\sqrt{\frac{k}{m}}\right) = f_{one}$$

The frequency with two identical springs is $\sqrt{2}$ times the frequency with one spring.
An even more general result can be obtained with similar reasoning even where the two k's differ, $f_{two}^2 = f_1^2 + f_2^2$.

P14.59. Prepare: The strategy will be to find how long the block on the left will take to fall to the table in free fall and use that as one-half of a period for the oscillator on the right. The spring is 30 cm long, so before release the spring is neither stretched nor compressed and the block is at its maximum height (not in equilibrium because the force of gravity is still acting on it). To go from maximum height to minimum height takes one-half of a period.

Once we know the period and the mass of the block on the right we can solve for k.

Solve:

$$\Delta y = \frac{1}{2} a_y (\Delta t)^2$$

$$\Delta t = \sqrt{\frac{2\Delta y}{a_y}} = \sqrt{\frac{2(-0.30 \text{ m})}{-9.8 \text{ m/s}^2}} = 0.247 \text{ s} = \frac{1}{2}T$$

$T = 2(0.247 \text{ s}) = 0.495 \text{ s}$.

Solve Equation 14.27 for k.

$$k = \frac{4\pi^2}{T^2} m = \frac{4\pi^2}{(0.495 \text{ s})^2} (0.050 \text{ kg}) = 8.1 \text{ N/m}$$

Assess: This is a reasonable spring, neither extremely stiff nor extremely loose.

The mass of the block on the left is irrelevant since all objects have the same acceleration in free fall.

The block on the right will pass through its equilibrium position when it falls far enough that $k\Delta y = mg$; that position is halfway down its total descent.

P14.61. Prepare: Because this is a "heavy bob supported on a very thin rod" we can model it as a simple pendulum; we will also assume the amplitude of oscillation is small.

As the temperature rises the thin steel rod will lengthen according to Equation 12.19, $\Delta L = \alpha L_i \Delta T$ (where this T is the temperature). Use Table 12.3 to look up $\alpha_{\text{steel}} = 12 \times 10^{-6} \text{ K}^{-1}$.

Solve: (a) The period of the simple pendulum is given by

$$T = 2\pi \sqrt{\frac{L}{g}} = 2\pi \sqrt{\frac{1.00000 \text{ m}}{9.80 \text{ m/s}^2}} = 2.00709 \text{ s}$$

(b) First compute the new length of the pendulum as follows:

$$\Delta L = \alpha L_i \Delta T = (12 \times 10^{-6} \text{ K}^{-1})(1.00000 \text{ m})(10 \text{ K}) = 0.00012 \text{ m}$$

$$L_f = L_i + \Delta L = 1.00012 \text{ m}$$

Now put the new length in the formula for the period.

$$T = 2\pi \sqrt{\frac{L}{g}} = 2\pi \sqrt{\frac{1.00012 \text{ m}}{9.80 \text{ m/s}^2}} = 2.00721 \text{ s}$$

(c) Every period the warm clock is off by the difference of the two periods, 0.00012 s; so the warm clock will be off by 1.0 s in 1/0.00012 s periods. That's 8304 periods, or $8304(2.00709 \text{ s}) = 16667 \text{ s} = 4.6 \text{ h}$.

Assess: When you do integrated problems like this you are even more likely than usual to run into conflicting uses of the same symbol. Please keep careful track of when T is the temperature and when it is the period.

Because of these timekeeping inaccuracies due to temperature variations, which are really not acceptable, clockmakers such as John Harrison in the 18th century made pendulums by combining strips of two different metals that counteracted each other's changes in length.

P14.63. Prepare: When we have two oscillators at slightly different frequencies we observe a regular pattern when they are in step and out of step. This is called beats.

Solve: **(a)** For the 24.8-cm pendulum

$$f = \frac{1}{2\pi}\sqrt{\frac{g}{L}} = \frac{1}{2\pi}\sqrt{\frac{9.8 \text{ m/s}^2}{0.248 \text{ m}}} = 1.00 \text{ Hz}$$

$$T = 2\pi\sqrt{\frac{L}{g}} = 2\pi\sqrt{\frac{0.248 \text{ m}}{9.8 \text{ m/s}^2}} = 1.00 \text{ s}$$

For the 38.8-cm pendulum

$$f = \frac{1}{2\pi}\sqrt{\frac{g}{L}} = \frac{1}{2\pi}\sqrt{\frac{9.8 \text{ m/s}^2}{0.388 \text{ m}}} = 0.800 \text{ Hz}$$

$$T = 2\pi\sqrt{\frac{L}{g}} = 2\pi\sqrt{\frac{0.388 \text{ m}}{9.8 \text{ m/s}^2}} = 1.25 \text{ s}$$

(b) Say they both tick at $t = 0.0$ s. The slower one will tick at $t = 1.25$ s, $t = 2.50$ s, $t = 3.75$ s, and $t = 5.0$ s, at which time it will be back on an integer second and coincide with the faster one, which ticks every second. It took 5.00 s for this pattern to happen and it will repeat each 5.00 s.
(c) The frequency of this phenomenon is $1/T = 1/5.00$ s $= 0.200$ Hz. What we notice is that this frequency is the difference of the original frequencies.
Assess: The beat frequency is the difference between the frequencies of the two oscillators. This phenomenon is most often observed in daily life with sound. Musicians tune their instruments by playing together, listening for the beats (which indicate the frequencies aren't identical), and adjusting their instruments or embouchures until the beats go away, or the beat frequency is zero.

P14.69. **Prepare:** The oscillator is in simple harmonic motion and is damped, so we will use Equation 14.33.
Solve: The maximum displacement, or amplitude, of a damped oscillator decreases as $x_{max}(t) = Ae^{-t/\tau}$, where τ is the time constant. We know $x_{max}/A = 0.60$ at $t = 50$ s, so we can find τ as follows:

$$-\frac{t}{\tau} = \ln\left(\frac{x_{max}(t)}{A}\right) \Rightarrow \tau = -\frac{50 \text{ s}}{\ln(0.60)} = 97.88 \text{ s}$$

Now we can find the time t_{30} at which $x_{max}/A = 0.30$

$$t_{30} = -\tau\ln\left(\frac{x_{max}(t)}{A}\right) = -(97.88 \text{ s})\ln(0.30) = 118 \text{ s}$$

The undamped oscillator has a frequency $f = 2$ Hz $= 2$ oscillations per second. Then the number of oscillations before the amplitude decays to 30% of its initial amplitude is $N = f \cdot t_{30} = (2 \text{ oscillations/s}) \cdot (118 \text{ s}) = 236$ oscillations or 240 oscillations to two significant figures.

P14.71. **Prepare:** In one time constant the displacement is decreased to 37% of its initial value, $y = A(1/e)$ $= .37A$; in two time constants it will be $A(1/e)^2 = 0.13A$.
The initial total energy is $E = \frac{1}{2}kA^2 = \frac{1}{2}(220 \text{ N/ms})(0.15 \text{ m})^2 = 2.475$ J.
Solve:

$$E_{new} = \frac{1}{2}ky^2 = \frac{1}{2}k(0.13A)^2 = (0.13)^2\frac{1}{2}kA^2 = (0.13)^2(2.475 \text{ J}) = 0.0453 \text{ J}$$

The energy dissipated is the difference between the original total energy and what is left. $E_{diss} = 2.475$ J $-$ 0.0453 J $= 2.43$ J, which should be rounded to 2.4 J.
Assess: There is really very little energy left after 2τ.

15

TRAVELING WAVES AND SOUND

Q15.1. Reason: **(a)** In a transverse wave, the thing or quantity that is oscillating, such as the particles in a string, oscillates in a direction that is transverse (perpendicular) to the direction of the propagation of the wave.
(b) Vibrations of a bass guitar string are a form of transverse wave. You can see that the oscillation is perpendicular to the string.
Assess: The plucking action makes the segment of the string move perpendicular to the string, but the disturbance is propagated along the string.

Q15.3. Reason: Wave speed is independent of wave amplitude, $v_1 = v_2 = v_3$.
Wave speed for mechanical waves depends on the properties of the medium, not the amplitude of the vibration.
Assess: If the wave speed were dependent on the amplitude then it might be the case that a later shout could overtake an earlier whisper.

Q15.7. Reason: The speed of sound in air depends on the temperature of the air. The distance across the stadium can be measured to give the path length of the sound. Then, you can take the path length and divide by the time between the emission and detection of the pulse to get the speed of the sound. Finally, once the speed of sound is known, you can find the corresponding temperature either by consulting a chart like Table 15.1 or by using Equation 15.3.
Assess: The advantage of measuring temperature this way is that it gives you an idea of the average temperature for the whole stadium. It also determines the temperature quickly—the time of measurement is simply the time it takes for sound to travel across the stadium.

Q15.9. Reason: Please refer to Figure Q15.9. Since the wave is traveling to the left, the snapshot has the same shape as the history graph. To understand why, consider that the history graph tells us about the displacement at one point in space. As the wave moves to the left, that point witnesses spots on the wave further to the right. Thus increasing t by Δt on the history graph has the same effect as increasing x by $v\Delta t$ on the snapshot graph, where v is the speed of the wave. Consequently the history graph and snapshot graph have the same shape. The only difference is that in going from the history graph to the snapshot graph, the horizontal axis is scaled by a factor of v.

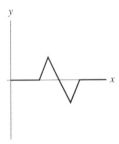

Assess: A similar argument can be used to show that if a wave is traveling to the *right*, the snapshot graph can be obtained from the history graph by reflecting the graph about the y-axis and scaling the horizontal axis by a factor of v.

Q15.11. Reason: The relationship between wavelength and frequency is $\lambda = v/f$. This tells us that the wavelength is inversely proportional to the frequency—the greater the frequency then, the smaller the wavelength. Since $f_1 < f_2 < f_3$, we can write $\lambda_1 > \lambda_2 > \lambda_3$.

Assess: Since the speed of sound is a constant for any given situation, the wavelength depends only on the frequency.

Q15.15. Reason:

$$\text{power} = \frac{\text{energy}}{\text{time}}$$

$$P_A = \frac{2 \text{ J}}{2 \text{ s}} = 1 \text{ W}$$

$$P_B = \frac{10 \text{ J}}{5 \text{ s}} = 2 \text{ W}$$

$$P_C = \frac{2 \text{ mJ}}{2 \text{ ms}} = 2 \text{ W}$$

$$P_B = P_C > P_A$$

Assess: We have seen this concept before; it is the ratio of energy to time that matters, not the energy number alone.

Problems

P15.1. Prepare: The wave is a traveling wave on a stretched string. We will use Equation 15.2 to find the wave speed.

Solve: The wave speed on a stretched string with linear density μ is $v_{string} = \sqrt{T_S/\mu}$. The wave speed if the tension is doubled will be

$$v'_{string} = \sqrt{\frac{2T_S}{\mu}} = \sqrt{2}\,v_{string} = \sqrt{2}\,(200 \text{ m/s}) = 280 \text{ m/s}$$

Assess: Wave speed increases with increasing tension.

P15.5. Prepare: Two pulses of sound are detected because one pulse travels through the metal to the microphone while the other travels through the air to the microphone. Sound travels *faster* through solids than gases, so the pulse traveling through the metal will reach the microphone *before* the pulse traveling through the air. We will take the speed of sound in air as 343 m/s.

Solve: The time interval for the sound pulse traveling through the air is

$$\Delta t_{air} = \frac{\Delta x}{v_{air}} = \frac{4.0 \text{ m}}{343 \text{ m/s}} = 0.01166 \text{ s} = 11.66 \text{ ms}$$

Because the pulses are separated in time by 11.0 ms, the pulse traveling through the metal takes $\Delta t_{metal} = 0.66$ ms to travel the 4.0 m to the microphone. Thus, the speed of sound in the metal is

$$v_{metal} = \frac{\Delta x}{\Delta t_{metal}} = \frac{4.0 \text{ m}}{0.00066 \text{ s}} = 6100 \text{ m/s}$$

Assess: We see from Table 15.1 that the speed obtained as shown is quite typical of metals.

P15.7. Prepare: Please refer to Figure P15.7. This is a wave traveling at constant speed. The pulse moves 1 m to the right every second.

Solve: The snapshot graph shows the wave at all points on the x-axis at $t = 0$ s. You can see that nothing is happening at $x = 6$ m at this instant of time because the wave has not yet reached this point. The leading edge of the wave is still 1 m away from $x = 6$ m. Because the wave is traveling at 1 m/s, it will take 1 s for the leading edge to reach $x = 6$ m. Thus, the history graph for $x = 6$ m will be zero until $t = 1$ s. The first part of the wave causes an upward displacement of the medium. The rising portion of the wave is 2 m wide, so it will take 2 s to pass the $x = 6$ m point. The constant part of the wave, whose width is 2 m, will take 2 seconds to pass $x = 6$ m and during this time the displacement of the medium will be a constant ($\Delta y = 1$ cm). The trailing edge of the pulse arrives at $t = 5$ s at $x = 6$ m. The displacement now becomes zero and stays zero for all later times.

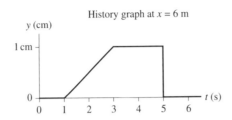

P15.9. Prepare: Please refer to Figure P15.9. This is a wave traveling at constant speed to the right at 1 m/s.
Solve: This is the history graph of the wave at $x = 0$ m. The graph shows that the $x = 0$ m point of the medium first sees the negative portion of the pulse wave at $t = 1$ s. Thus, the snapshot graph of this wave at $t = 1$ s must have the leading negative portion of the wave at $x = 0$ m.

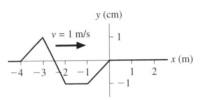

Snapshot graph at $t = 1.0$ s

P15.11. Prepare: We will use Equation 15.9 to find the wave speed.
Solve: The wave speed is

$$v = \frac{\lambda}{T} = \frac{2.0 \text{ m}}{0.20 \text{ s}} = 10 \text{ m/s}$$

P15.15. Prepare: The wave is a traveling wave. A comparison of the given wave equation with Equation 15.8 yields $A = 5.2$ cm, $2\pi/\lambda = 5.5$ rad/m, and $2\pi/T = 72$ rad/s.
Solve: (a) The frequency is

$$f = \frac{1}{T} = \left(\frac{1}{2\pi}\right)\frac{2\pi}{T} = \frac{72 \text{ rad/s}}{2\pi} = 11 \text{ Hz}$$

to two significant figures.
(b) The wavelength is

$$\frac{2\pi}{\lambda} = 5.5 \text{ rad/m} \Rightarrow \lambda = \frac{2\pi}{5.5 \text{ rad/m}} = 1.1 \text{ m}$$

to two significant figures.
(c) The wave speed $v = \lambda f = 13$ m/s to two significant figures.

P15.17. Prepare: Please refer to Figure P15.17.
Solve: A visual examination of the graph shows that the maximum displacement of the wave, called the amplitude, is 4.0 cm.
The wavelength is the distance between two consecutive peaks, which gives $\lambda = 14$ m $- 2$ m $= 12$ m.
The frequency of the wave using Equation 15.10 is

$$f = \frac{v}{\lambda} = \frac{24 \text{ m/s}}{12 \text{ m}} = 2.0 \text{ Hz}$$

Assess: It is important to know how to read information from graphs.

P15.21. Prepare: We will use the fundamental relation for periodic waves in Equation 15.10 to solve for the wavelength. We must first look up the speed of sound in water; Table 15.1 says it is 1480 m/s.
Solve:

$$\lambda = \frac{v}{f} = \frac{1480 \text{ m/s}}{100 \times 10^3 \text{ Hz}} = 15 \text{ mm} = 1.5 \text{ cm}$$

Assess: Because the speed of sound in water is over four times the speed of sound in air, dolphins must be quick to process the sonar information.

P15.27. Prepare: The power (or energy/time) is the intensity multiplied by the area. The intensity is $I = 1.0 \times 10^{-6}$ W/m^2. We can deduce from the information given that the area of the eardrum is $A = \pi R^2 = \pi (4.2 \text{ mm})^2 = 5.54 \times 10^{-5} \text{ m}^2$.
Solve:

$$P = Ia = (1.0 \times 10^{-6} \text{ W/m}^2)(5.54 \times 10^{-5} \text{ m}^2) = 5.5 \times 10^{-11} \text{ W}$$

The energy delivered to your eardrum each second is 55 pJ.
Assess: This is an incredibly tiny amount of energy per second; you should be impressed that your ear can detect such small signals! Your eardrum moves back and forth about the width of 100 atoms in such cases!

P15.29. Prepare: We are asked to find the energy received by your back of area 30 cm × 50 cm in 1.0 h if the electromagnetic wave intensity is 1.4×10^3 W/m^2. The energy delivered to your back in time t is $E = Pt$, where P is the power of the electromagnetic wave. The intensity of the wave is $I = P/a$ where a is the area of your back.
Solve: The energy received by your back is

$$E = Pt = Iat = (0.80)(1400 \text{ W/m}^2)(0.30 \times 0.50 \text{ m}^2)(3600 \text{ s}) = 6.1 \times 10^5 \text{ J}$$

Assess: This is equivalent to receiving approximately 170 J of energy per second by your back. This energy is relatively large and will certainly lead to tanning.

P15.33. Prepare: To find the power of a laser pulse, we need the energy it contains, U, and the time duration of the pulse, Δt. Then to find the intensity, we need the area of the pulse. Its radius is 0.50 mm.
Solve: (a) Using $P = U / \Delta t$, we find the following:

$$P = (1.0 \times 10^{-3} \text{ J})/(15 \times 10^{-9} \text{ s}) = 6.67 \times 10^4 \text{ W}$$

(b) Then from $I = P / a$, we obtain

$$I = \frac{(6.67 \times 10^4 \text{ W})}{\pi (5.0 \times 10^{-4} \text{ m})^2} = 8.5 \times 10^{10} \text{ W/m}^2$$

Assess: This is very intense light. Using the data from Problem 15.32, the laser light is about 400 million times as intense as the energy from the sun.

P15.35. Prepare: Table 15.3 tells us that the intensity of a whisper at one meter is $I_1 = 1.0 \times 10^{-10}$ W/m^2. We'll use Equation 15.12 and ratios to find what it would be twice as far away, $I = \dfrac{P_{\text{source}}}{4\pi r^2}$.

Solve:

$$\frac{I_2}{I_1} = \frac{\dfrac{P_{source}}{4\pi r_2^2}}{\dfrac{P_{source}}{4\pi r_1^2}} = \frac{r_1^2}{r_2^2} = \frac{(1.0\text{ m})^2}{(2.0\text{ m})^2} = 0.25$$

$$I_2 = (0.25)I_1 = (0.25)(1.0\times10^{-10}\text{ W/m}^2) = 0.25\times10^{-10}\text{ W/m}^2 = 2.5\times10^{-11}\text{ W/m}^2$$

The sound intensity level is given by Equation 15.14 where $I_0 = 1.0\times10^{-12}\text{ W/m}^2$.

$$\beta = (10\text{ dB})\log_{10}\left(\frac{I}{I_0}\right) = (10\text{ dB})\log_{10}\left(\frac{2.5\times10^{-11}\text{ W/m}^2}{1.0\times10^{-12}\text{ W/m}^2}\right) = 14\text{ dB}$$

Assess: Table 15.3 gives the sound intensity level for a whisper at 1 m as 20 dB; we expect it to be less at 2 m, and 14 dB is just about what we expect.

P15.39. Prepare: Knowing the sound intensity level $\beta_1 = 120\text{ dB}$ at the point $r_1 = 5\text{ m}$, we can determine the intensity I_1 at this point. Since the power output of the speaker is a constant, knowing the intensity I_1 and distance r_1 from the speaker at one point, we can determine the intensity I_2 at a second point r_2. Knowing the intensity of the sound I_2 at this second point r_2, we can determine the sound intensity level β_2 at the second point.
Solve: The sound intensity may be determined from the sound intensity level as follows: $\beta_1 = (10\text{ dB})\log_{10}(I_1/I_o)$.
Inserting numbers: $120\text{ dB} = (10\text{ dB})\log_{10}[I_1/(10^{-12}\text{ W/m}^2)]$ or $12 = \log_{10}[I_1/(10^{-12}W/m^2)]$

Taking the antilog of both sides obtains $10^{12} = I_1/(10^{-12}\text{ W/m}^2)$ which may be solved for I_1 to obtain $I_1 = (10^{-12}\text{ W/m}^2)$ $(10^{12}) = 1\text{ W/m}^2$.
Knowing how the sound intensity is related to the power and that the power does not change, we can determine the sound intensity at any other point.
Inserting $a = 4\pi r^2$, into $P = I_1 a_1 = I_2 a_2$ obtain $I_2 = I_1(r_1/r_2)^2 = (1\text{ W/m}^2)(5/35)^2 = 2.04\times10^{-2}\text{ W/m}^2$.
Finally, knowing the sound intensity at the second point r_2, we can determine the sound intensity level at this point by

$$\beta_2 = (10\text{ dB})\log_{10}(I_2/I_o) = (10\text{ dB})\log_{10}(2.04\times10^{-2}/1\times10^{-12}) = (10\text{ dB})\log_{10}(2.04\times10^{10})$$

continuing

$$\beta_2 = 10\text{ dB}[\log_{10}(2.04) + \log_{10}(10^{10})] = 10\text{ dB}[0.31+10] = 103\text{ dB}$$

Assess: This is still as loud as a pneumatic hammer. Either take ear protection or move farther away from the speaker.

P15.41. Prepare: The frequency of the opera singer's note is altered by the Doppler effect. The frequency is f_+ as the car approaches and f_- as it moves away. f_o is the frequency of the source. The speed of sound in air is 343 m/s.
Solve: (a) Using 90 km/hr = 25 m/s, the frequency as her convertible approaches the stationary person is

$$f_+ = \frac{f_0}{1 - v_S/v} = \frac{600\text{ Hz}}{1 - \dfrac{25\text{ m/s}}{343\text{ m/s}}} = 650\text{ Hz}$$

(b) The frequency as her convertible recedes from the stationary person is

$$f_- = \frac{f_0}{1+v_s/v} = \frac{600 \text{ Hz}}{1+\dfrac{25 \text{ m/s}}{343 \text{ m/s}}} = 560 \text{ Hz}$$

Assess: As would have been expected, the pitch is higher in front of the source than it is behind the source.

P15.45. Prepare: First compute $f_0 = 3.0/10 \text{ s} = 0.30 \text{ Hz}$.

When the detector (you in the boat) is moving, the measured frequency is $f_- = 2.0/10 \text{ s} = 0.20 \text{ Hz}$.

The problem gives $v_o = 1.5 \text{ m/s}$.

Use Equation 15.17, $f_- = (1 - v_o/v) f_0$, and solve for v, the speed of the waves.

Solve:

$$\frac{f_-}{f_0} = 1 - \frac{v_o}{v}$$

$$\frac{v_o}{v} = 1 - \frac{f_-}{f_0}$$

$$v = v_o \left(\frac{1}{1 - \dfrac{f_-}{f_0}} \right)$$

$$v = v_o \left(\frac{f_0}{f_0 - f_-} \right)$$

$$v = (1.5 \text{ m/s}) \left(\frac{0.30 \text{ Hz}}{0.30 \text{ Hz} - 0.20 \text{ Hz}} \right) = 4.5 \text{ m/s}$$

Assess: These are pretty fast water waves, but within reason. The speed of surface water waves depends on the depth of the water, but is typically in the range of 2–3 m/s. However, tsunamis can travel much faster than this. Carefully note the subscripts: The o stands for observer while the 0 stands for the initial or original (frequency). They look similar but are different.

P15.49. Prepare: The explosive's sound travels down the lake and into the granite, and then it is reflected by the oil surface. A pictorial representation of the problem along with the speed of sound in water and granite is shown.

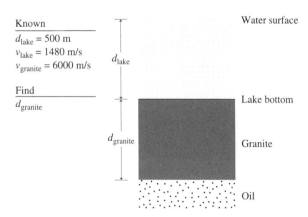

Solve: The echo time is equal to

$$t_{echo} = t_{water\ down} + t_{granite\ down} + t_{granite\ up} + t_{water\ up}$$

$$0.94\ s = \frac{500\ m}{1480\ m/s} + \frac{d_{granite}}{6000\ m/s} + \frac{d_{granite}}{6000\ m/s} + \frac{500\ m}{1480\ m/s} \Rightarrow d_{granite} = 793\ m$$

Assess: Drilling into granite for 3/4 km is not unreasonable.

P15.53. Prepare: The wave pulse is a traveling wave on a stretched string. The pulse travels along the rope with speed $v = \Delta x / \Delta t = 3.0\ m/150\ s = 20\ m/s = \sqrt{T_s/\mu}$, where we used the relationship involving the wave's speed, the rope's tension, and the linear density $\mu = m_{rope}/L$. We can find the tension in the rope from a Newtonian analysis. Bob experiences horizontal forces $\vec{T}$ and $\vec{f}_k$. He is being pulled at constant speed, so he is in dynamic equilibrium with $\vec{F}_{net} = 0$ N. Thus $T = f_k = \mu_k n$. From the vertical forces we see that $n = w = m_{Bob}g$.

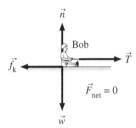

Solve: The rope's tension is

$$T_s = \mu_k m_{Bob} g = (0.20)(60\ kg)(9.8\ m/s^2) = 117.6\ N$$

Using this in the velocity equation, we can find the rope's mass as follows:

$$v = \sqrt{\frac{T_s}{m_{rope}/L}} \Rightarrow m_{rope} = \frac{LT_s}{v^2} = \frac{(3.0\ m)(117.6\ N)}{(20\ m/s)^2} = 0.882\ kg = 880\ g$$

to two significant figures.
Assess: The mass of 880 g for a 3.0-m-long rope is reasonable.

P15.55. Prepare: Please refer to Figure P15.55. The wave pulse is a traveling wave on a stretched string. While the tension T_s is the same in both the strings, the wave speeds in the two strings are not. We have

$$v_1 = \sqrt{\frac{T_s}{\mu_1}} \quad \text{and} \quad v_2 = \sqrt{\frac{T_s}{\mu_2}} \Rightarrow v_1^2 \mu_1 = v_2^2 \mu_2 = T_s$$

Because $v_1 = L_1/t_1$ and $v_2 = L_2/t_2$, and because the pulses are to reach the ends of the string simultaneously the previous equation can be simplified to

$$\frac{L_1^2 \mu_1}{t^2} = \frac{L_2^2 \mu_2}{t^2} \Rightarrow \frac{L_1}{L_2} = \sqrt{\frac{\mu_2}{\mu_1}}$$

Solve:

$$\frac{L_1}{L_2} = \sqrt{\frac{\mu_2}{\mu_1}} = \sqrt{\frac{4.0\ g/m}{2.0\ g/m}} = \sqrt{2} \Rightarrow L_1 = \sqrt{2}L_2$$

Since $L_1 + L_2 = 4$ m,

$$\sqrt{2}L_2 + L_2 = 4\ m \Rightarrow L_2 = 1.66\ m \quad \text{and} \quad L_1 = \sqrt{2}(1.66\ m) = 2.34\ m$$

The two lengths will be reported as 1.7 m and 2.3 m.

P15.57. Prepare: Draw a right triangle with the base 3.0 m long and the height 0.15 m long. We will keep some extra significant figures to see the point of this problem.
We are given $d_R = 3.0$ m, the distance between the ears is 0.15 m, and we use $v = 343$ m/s.
Solve: (a) Use the Pythagorean theorem to compute the distance from the bird to the coyote's left ear.

$$d_L = \sqrt{(3.0\text{ m})^2 + (0.15\text{ m})^2} = 3.003748\text{ m}$$

To two significant figures this is 3.0 m, but we need the extra digits in this case to get a nonzero answer.
(b) Use $t = d/v$.

$$t_L - t_R = \frac{d_L}{v} - \frac{d_R}{v} = \frac{d_L - d_R}{v} = \frac{3.003748\text{ m} - 3.0\text{ m}}{343\text{ m/s}} = 1.1 \times 10^{-5}\text{ s} = 11\ \mu s$$

(c) The period of the sound wave is $T = 1/f = 1 \times 10^{-3}$ s. The requested ratio is

$$\frac{t_L - t_R}{T} = \frac{1.1 \times 10^{-5}\text{ s}}{1 \times 10^{-3}\text{ s}} = 1.1 \times 10^{-2} = 0.011$$

Assess: This is impressive time resolution, even down to a hundredth of a period of the wave.

P15.61. Prepare: We'll employ the fundamental relation for periodic waves, $v = \lambda f$.
Solve: (a) We simply want to know the period of the passing waves, $T = 1/f = 1/0.3\text{ Hz} = 3.33\text{ s} \approx 3\text{ s}$.
(b) We are given that $\lambda = 30$ m and $v = \lambda f = (30\text{ m})(0.3\text{ Hz}) = 9$ m/s.
Assess: We have only one significant figure for the data in our estimation.
Horizontal oscillations can also contribute to motion sickness, as can rolling oscillations. In fact, motion sickness can even be produced without physical oscillation of the body, but by moving (including oscillating) visual stimuli such as flight trainers.

P15.63. Prepare: All quantities, except the period, needed to write the y-equation for a wave traveling in the negative x-direction are given. We will determine the period using Equation 15.9. The y-equation that we are asked to write will look like Equation 15.8.
Solve: The period is calculated as follows: $T = \dfrac{\lambda}{v} = \dfrac{0.50\text{ m}}{4.0\text{ m/s}} = 0.125$ s.

The displacement equation for the wave is

$$y(x, t) = (5.0\text{ cm})\cos\left[2\pi\left(\frac{x}{50\text{ cm}} + \frac{t}{0.125\text{ s}}\right)\right]$$

Assess: The positive sign in the cosine function's argument indicates motion along the $-x$ direction.

P15.65. Prepare: The function of x and t with the minus sign will be a wave traveling to the right.
Solve: (a) For simplicity, take the snapshot at $t = 0$ s so the equation becomes $y(x) = (3.0\text{ cm})\cos(1.5x)$.

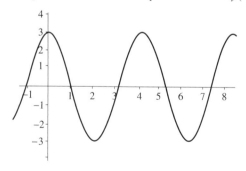

(b) First compare Equation 15.7 for a wave moving to the right to our wave function to identify λ and T.

$$y(x,t) = A\cos\left[2\pi\left(\frac{x}{\lambda} - \frac{t}{T}\right)\right]$$

$$y(x,t) = (3.0\text{ cm})\cos(1.5x - 50t)$$

We see that $\lambda = (2\pi/1.5)$ m, $T = (2\pi/50)$ s, and (unneeded) $A = 3.0$ cm.

Now use Equation 15.9 for the wave speed.

$$v = \frac{\lambda}{T} = \frac{(2\pi/1.5)\text{ m}}{(2\pi/50)\text{ s}} = 33\text{ m/s}$$

As mentioned, because of the minus sign in Equation 15.7 (and our wave function), the wave is moving to the right.

Assess: We do not know what kind of wave this is, but the answer is reasonable for the wave speed on a string, and the amplitude is a reasonable amplitude for a wave on a string. Note that although we had to concern ourselves with the 2π, it did cancel out in our last calculation.

P15.69. Prepare: The radio wave is an electromagnetic wave. At a distance r, the 25 kW power station spreads out waves to cover the surface of a sphere of radius r. The surface area of a sphere is $4\pi r^2$.

Solve: Thus, the intensity of the radio waves is

$$I = \frac{P_{\text{source}}}{4\pi r^2} = \frac{25\times 10^3\text{ W}}{4\pi(10\times 10^3\text{ m})^2} = 2.0\times 10^{-5}\text{ W/m}^2$$

Assess: The frequency of the radio waves does not enter the calculations above.

P15.75. Prepare: When the problem says "radiated equally in all directions" we know that it will be a uniform spherical wave, and that we can use Equation 15.12.

$$I = \frac{P_{\text{source}}}{4\pi r^2}$$

We are given that $P_{\text{source}} = 600$ W and $r = 5.00$ m.

Solve: (a)

$$I = \frac{P_{\text{source}}}{4\pi r^2} = \frac{600\text{ W}}{4\pi(5.00\text{ m})^2} = 1.9\text{ W/m}^2$$

(b) Now we need Equation 15.14 for the sound intensity level where $I_0 = 1.0\times 10^{-12}\text{ W/m}^2$.

$$\beta = (10\text{ dB})\log_{10}\left(\frac{I}{I_0}\right) = (10\text{ dB})\log_{10}\left(\frac{1.9\text{ W/m}^2}{1.0\times 10^{-12}\text{ W/m}^2}\right) = 122.8\text{ dB} \approx 123\text{ dB}$$

(c) We have to do the previous calculations in reverse. 23 dB less than 123 dB is 100 dB, so we plug $\beta = 100$ dB into Equation 15.15.

$$I = (1.0\times 10^{-12}\text{ W/m}^2)10^{(\beta/10\,\text{dB})} = (1.0\times 10^{-12}\text{ W/m}^2)10^{(100\,\text{dB}/10\,\text{dB})} = 0.010\text{ W/m}^2$$

The other part of the reverse calculation (with the new intensity of 0.010 W/m^2) is to solve the first equation for r.

$$r = \sqrt{\frac{P_{\text{source}}}{4\pi I}} = \sqrt{\frac{600\text{ W}}{4\pi(0.010\text{ W/m}^2)}} = 69\text{ m}$$

Phil needs to be 14 times farther away than you.

Assess: And at 100 dB he still shouldn't have any trouble hearing the concert (and neither will you with the earplugs at 5.00 m).

P15.77. Prepare: The sound generator's frequency is altered by the Doppler effect. According to Equations 15.16, the frequency increases as the generator approaches the student, and it decreases as the generator recedes from the student. Convert rpm into SI units. Use 343 m/s for the speed of sound.
Solve: The generator's speed is

$$v_{\text{s}} = r\omega = r(2\pi f) = (1.0 \text{ m})2\pi\left(\frac{100}{60} \text{ rev/s}\right) = 10.47 \text{ m/s}$$

The frequency of the approaching generator is

$$f_{+} = \frac{f_0}{1 - v_{\text{s}}/v} = \frac{600 \text{ Hz}}{1 - \frac{10.47 \text{ m/s}}{343 \text{ m/s}}} = 620 \text{ Hz}$$

Doppler effect for the receding generator, on the other hand, is

$$f_{-} = \frac{f_0}{1 + v_{\text{s}}/v} = \frac{600 \text{ Hz}}{1 + \frac{10.47 \text{ m/s}}{343 \text{ m/s}}} = 580 \text{ Hz}$$

Thus, the highest and the lowest frequencies heard by the student are 620 Hz and 580 Hz.

SUPERPOSITION AND STANDING WAVES

Q16.1. Reason: Where there is a change in medium—in particular a change in the wave speed—then reflection can occur.
Assess: Light travels at different speeds in water and air, and so some is reflected at a water-air interface.

Q16.3. Reason: See Figure 16.10 in the text.
(a) In Chapter 15 we saw that the speed of a wave on a stretched string is $v = \sqrt{T_s/\mu}$. Since the left side of the string has a lower speed, the linear density must be greater there.
(b) You would start a pulse from the left side, in the part with the greater linear density, in order to have the reflection not inverted.
Assess: This would be a fairly easy experiment to set up at home with two different strings tied together.

Q16.9. Reason: The advantage of having low-frequency organ pipes closed at one end is that they will sound an octave lower without being twice as long as an open-open pipe. Because a pipe closed on one end contains only 1/4 of a wavelength in the lowest mode, the wavelength is twice as long as for an open-open pipe of the same length. If the wavelength is twice as long then the frequency is half, and that corresponds to a musical interval of an octave.
Assess: Pipes closed on one end also have a different sound (timbre) than open-open pipes sounding the same note.

Q16.11. Reason: While the fundamental frequencies of normal voices are below 3000 Hz, there are higher harmonics mixed in that give the voice its characteristic sound and convey information. These harmonics are lost when the telephone cuts off frequencies above 3000 Hz, and it is a bit harder to understand what is said.
Assess: The range of human hearing is usually quoted as 20 Hz–20 kHz; a large portion of that range is above 3000 Hz. This principle holds for music as well as spoken word. CDs are designed to reproduce frequencies clearly up to 20 kHz (and even a tiny bit higher), even though the musical notes themselves (the fundamental frequencies) are not nearly that high. Those higher harmonics give the sound definition and sharpness; the music sounds muddled if the high frequencies are cut out.

Q16.15. Reason: It is the harmonics and formants that help us understand what is being said in the first place (that is, the harmonics of the vocal cords, modified by the formants of the vocal tract, make an "ee" sound different from an "oo" sound at the same pitch). The harmonics are multiples of the fundamental, and when the fundamental is high (1000 Hz or more), the harmonics are out of the range of hearing for many people. The formants emphasize different harmonics and allow us to distinguish different vocalizations, but if the harmonics cannot be heard then it would be difficult to tell the difference between an "ee" sound and an "oo" sound.
Assess: The upper limit on the frequencies people can hear varies from person to person. Men and people who have endured lengthy periods of very loud sounds cannot hear frequencies as high as women and people who've protected their hearing, on average.

Q16.17. Reason: The quality of the sound of your voice depends on the resonant frequencies of the cavities (throat, mouth, and nose) in your vocal tract. When one of these cavities is stuffed up, the quality of your voice will be affected.

Assess: In this case the fundamental frequency has not changed but the ability to create and enhance certain harmonics has. The harmonics play an important role in the quality of the sound of your voice.

Q16.19. Reason: Two points that experience crests at the same time would have their displacements in phase. The displacements at points A, B, C, D, and F are all experiencing crests at the same time. The displacement at point E is has a trough at that time.

(a) Displacements at A and B are in phase; they experience the *same* crest at the same time.

(b) Displacements at C and D are in phase; they experience different crests at the same time.

(c) Displacements at E and F are out of phase; E experiences a trough at the times that F experiences a crest.

Assess: Half a period later the displacements at points A, B, C, D, and F are all troughs, but they are still in phase. The displacement at point E will always be exactly out of phase with the displacements at all the other labeled points.

Problems

P16.1. Prepare: The principle of superposition comes into play whenever the waves overlap. The waves are approaching each other at a speed of 1 m/s, that is, each part of each wave is moving 1 m every second.

Solve: The graph at $t = 1$ s differs from the graph at $t = 0$ s in that the left wave has moved to the right by 1 m and the right wave has moved to the left by 1 m. This is because the distance covered by the wave pulse in 1 s is 1 m. The snapshot graphs at $t = 2$ s, 3 s, and 4 s are a superposition of the left- and the right-moving waves. The overlapping parts of the two waves are shown by the dotted lines.

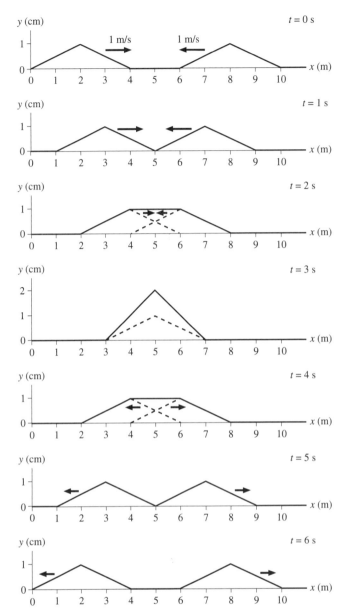

Assess: This is an excellent problem because it allows you to see the progress of each wave and the superposition (addition) of the waves. As time progresses, you know exactly what has happened to each wave and to the superposition of these waves.

P16.3. Prepare: The principle of superposition comes into play whenever the waves overlap.
Solve: As graphically illustrated in the following figure, the snapshot graph of Figure P16.3 was taken at $t = 4$ s.

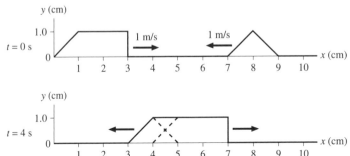

Assess: This is an excellent problem because it allows you to see the progress of each wave and the superposition (addition) of the waves. As time progresses, you know exactly what has happened to each wave and to the superposition of these waves.

P16.7. Prepare: Reflections at both ends of the string cause the formation of a standing wave. Figure P16.7 indicates that there are three full wavelengths on the 2.0-m-long string and that the wave speed is 40 m/s. We will use Equation 15.10 to find the frequency of the standing wave.

Solve: The wavelength of the standing wave is $\lambda = \frac{1}{3}(2.0 \text{ m}) = 0.667$ m. The frequency of the standing wave is thus

$$f = \frac{v}{\lambda} = \frac{40 \text{ m/s}}{0.667 \text{ m}} = 60 \text{ Hz}$$

Assess: The units are correct and this is a reasonable frequency for a vibrating string.

P16.9. Prepare: We assume that the string is tied down at both ends so there are nodes there. This means the length of the string is $L = \frac{1}{2}\lambda$ in the fundamental mode (there are no nodes between the ends). $\lambda = 2L = 2(0.89 \text{ m}) = 1.78$ m.

Solve: Use the fundamental relationship for periodic waves: $v = \lambda f = (1.78 \text{ m})(30 \text{ Hz}) = 53$ m/s.

Assess: Remember, we are talking about the speed of the wave on the string, not the speed of sound in air. These numbers are reasonable for bass guitar strings.

P16.11. Prepare: A string fixed at both ends supports standing waves. A standing wave can exist on the string only if its wavelength is given by Equation 16.1, that is, $\lambda_m = \frac{2L}{m}$, $m = 1, 2, 3\ldots$ The length L of the string is 240 cm.

Solve: (a) The three longest wavelengths for standing waves will therefore correspond to $m = 1, 2$, and 3. Thus,

$$\lambda_1 = \frac{2(2.40 \text{ m})}{1} = 4.80 \text{ m} \qquad \lambda_2 = \frac{2(2.40 \text{ m})}{2} = 2.40 \text{ m} \qquad \lambda_3 = \frac{2(2.40 \text{ m})}{3} = 1.60 \text{ m}$$

(b) Because the wave speed on the string is unchanged from one m value to the other,

$$f_2\lambda_2 = f_3\lambda_3 \Rightarrow f_3 = \frac{f_2\lambda_2}{\lambda_3} = \frac{(50.0 \text{ Hz})(2.40 \text{ m})}{1.60 \text{ m}} = 75.0 \text{ Hz}$$

Assess: The units on each determination are correct and the values are reasonable. The maximum wavelength of a standing wave in a string is twice the length of the string and all other possible wavelengths are fractions of this value.

P16.17. Prepare: We want to use the fundamental relationship for periodic waves. But first, convert the length to SI units (use Table 1.3).

$$L = 18 \text{ ft}\left(\frac{0.30 \text{ m}}{1 \text{ ft}}\right) = 5.49 \text{ m}$$

Solve: (a) For an open-closed pipe in the fundamental mode $L = \frac{1}{4}\lambda$ (one-quarter of a wavelength fits in the pipe),

$$\lambda = 4L = 4(5.49 \text{ m}) = 21.9 \text{ m}$$

$$f = \frac{v}{\lambda} = \frac{350 \text{ m/s}}{21.9 \text{ m}} = 16 \text{ Hz}$$

This is below the arbitrary lower limit of the range of human hearing.

(b) We notice that the true value (27.5 Hz) is different from the answer we got in part **(a)**. To find the "effective length" of the instrument with a fundamental frequency of 27.5 Hz using the open-closed tube model, we simply do the previous calculations in reverse order. First find λ for the fundamental mode.

$$\lambda = \frac{v}{f} = \frac{350 \text{ m/s}}{27.5 \text{ Hz}} = 12.7 \text{ m}$$

Now

$$L = \frac{1}{4} \lambda = \frac{1}{4} (12.7 \text{ m}) = 3.18 \text{ m} = 10.4 \text{ ft} \approx 10 \text{ ft}$$

Assess: The "effective length" is just over half of the real length. This would lead us to conclude that the open-closed tube model is not a very accurate model for this situation. A good contrabassoon gives foundation and body to the orchestra.

P16.19. Prepare: For the open-open tube, the two open ends exhibit antinodes of a standing wave. The possible wavelengths for this case are $\lambda_m = 2L/m$, where $m = 1, 2, 3 \ldots$ On the other hand, in the case of an open-closed tube $\lambda_m = 4L/m$, where $m = 1, 3, 5 \ldots$ The length of the tube is 121 cm.

Solve: (a) The three longest wavelengths are

$$\lambda_1 = \frac{2(1.21 \text{ m})}{1} = 2.42 \text{ m} \qquad \lambda_2 = \frac{2(1.21 \text{ m})}{2} = 1.21 \text{ m} \qquad \lambda_3 = \frac{2(1.21 \text{ m})}{3} = 0.807 \text{ m}$$

(b) The three longest wavelengths are

$$\lambda_1 = \frac{4(1.21 \text{ m})}{1} = 4.84 \text{ m} \qquad \lambda_2 = \frac{4(1.21 \text{ m})}{3} = 1.61 \text{ m} \qquad \lambda_3 = \frac{4(1.21 \text{ m})}{5} = 0.968 \text{ m}$$

Assess: It is clear that the end of the air column, whether open or closed, changes the possible modes.

P16.27. Prepare: First let's agree that the temperature of the air in the vocal tract is a little over 20°C and hence the speed of sound is 350 m/s. The relationship between speed, wavelength, and frequency for a traveling wave disturbance in any medium is $v = f\lambda$. The frequency of vibration in air is caused by and is the same as the frequency of vibration of the vocal cords. The length of the vocal tract is an integral number of half-wavelengths $L = m\lambda / 2$. The length of the vocal tract and hence the wavelengths that cause standing wave resonance do not change as the diver descends. However, since the speed of the sound waves changes, the frequency will also change.

Solve: When a sound of frequency 270 Hz is coming out of the vocal tract, the wavelength of standing waves established in the vocal tract associated with this frequency is

$$\lambda = v / f = (350 \text{ m/s}) / (270 \text{ Hz}) = 1.296 \text{ m}$$

As the diver descends, the vocal tract does not change length and hence this wavelength for standing wave resonance will not change. However since the sound is now travelling through a helium-oxygen mixture with a speed of 750 m/s, the frequency of the sound will change to

$$f = v / \lambda = (750 \text{ m/s}) / (1.296 \text{ m}) = 580 \text{ Hz}$$

Going through the same procedure for sound at a frequence of 2300 Hz we get the frequency in the helium-oxygen mixture to be $f = 4900$ Hz.

Assess: We are aware that the sound should be at a higher frequency and the frequencies obtained have higher values.

P16.29. Prepare: Interference occurs as a result of the path difference between the path lengths of the sound from the two speakers. A separation of 20 cm between the speakers leads to maximum intensity on the x-axis, but a separation of 30 cm leads to zero intensity.

Solve: (a) When the waves are in phase and lead to constructive interference, $(\Delta d)_1 = m\lambda = 20$ cm. For destructive interference, $(\Delta d)_2 = (m + \frac{1}{2}) \lambda = 30$ cm. Thus, for the same value of m

$$(\Delta d)_2 - (\Delta d)_1 = \frac{\lambda}{2} \Rightarrow \lambda = 2(30 \text{ cm} - 20 \text{ cm}) = 40 \text{ cm}$$

(b) If the distance between the speakers continues to increase, the intensity will again be a maximum when the separation between the speakers that produced a maximum has increased by one wavelength. That is, when the separation between the speakers is 20 cm + 40 cm = 60 cm.

Assess: The distances obtained are reasonable. As a check on our work we might want to determine the frequency of sound associated with the wavelength. A wavelength of 40 cm is associated with a frequency of 860 Hz, which is in the audible range.

P16.31. Prepare: Please refer to Figure P16.31. The circular wave fronts emitted by the two sources show that the two sources are in phase. This is because the wave fronts of each source have moved the same distance from their sources.

Solve: Let us label the top source as 1 and the bottom source as 2. For the point P, $d_1 = 3\lambda$ and $d_2 = 4\lambda$. Thus, $\Delta d = d_2 - d_1 = 4\lambda - 3\lambda = \lambda$. This corresponds to constructive interference.

For the point Q, $d_1 = \frac{7}{2}\lambda$ and $d_2 = 2\lambda$, so $\Delta d = |d_2 - d_1| = |2\lambda - 7\lambda/2| = 3\lambda/2$. This corresponds to destructive interference.

For the point R, $d_1 = \frac{5}{2}\lambda$ and $d_2 = \frac{7}{2}\lambda$, $\Delta d = d_2 - d_1 = \lambda$. This corresponds to constructive interference.

	r_1	r_2	Δr	C/D
P	3λ	4λ	λ	C
Q	$\frac{7}{2}\lambda$	2λ	$\frac{3}{2}\lambda$	D
R	$\frac{5}{2}\lambda$	$\frac{7}{2}\lambda$	λ	C

Assess: When the path difference Δr is an integral number of whole wavelengths (cases P and R), constructive interference occurs. When the path difference Δr is an integral number of half wavelengths (case Q), destructive interference occurs.

P16.33. Prepare: Knowing the following relationships for a vibrating string, $v = \sqrt{T/\mu}$, $v = f\lambda$, and $L = m\lambda/2$, we can establish what happens to the frequency as the tension is increased. Knowing the relationship $f_{\text{beat}} = f_1 - f_2$, we can determine the frequency of the string with the increased tension.

Solve: Combining the first three expressions above, obtain $f = \dfrac{m}{2L}\sqrt{\dfrac{T}{\mu}}$, which allows us to determine that the frequency will increase as the tension increases. Using the expression for the beat frequency, we know that the difference frequency between the two frequencies is 3 Hz. Combining this with our knowledge that the frequency increases, we obtain

$$f_{\text{beat}} = f_1 - f_2 \Rightarrow 3\,\text{Hz} = f_1 - 200\,\text{Hz} \Rightarrow f_1 = 203\,\text{Hz}$$

Assess: f_1 is larger than f_2 because the increased tension increases the wave speed and hence the frequency.

P16.35. Prepare: We'll use the fundamental relationship for periodic waves ($v = \lambda f$, where $f = 440$ Hz) and we'll compute λ from Equation 16.1 ($\lambda_1 = 2L = 2.0$ m, where $m = 1$) because the problem is about the fundamental frequency.

Solve:

$$v = \lambda f = (2.0\,\text{m})(440\,\text{Hz}) = 880\,\text{m/s}$$

Assess: This speed is over twice the speed of sound in air, but this string might be thin (small μ) and under a lot of tension in order to make the wave speed that high.

P16.37. Prepare: According to Equation 16.6, the standing wave on a guitar string, vibrating at its fundamental frequency, has a wavelength λ equal to twice the length L. We will first calculate the frequency of the wave that the string produces using Equations 15.2 and 15.10. The wave created by the guitar string travels as a sound wave with the same frequency but with a speed of 343 m/s in air.

Solve: The wave speed on the stretched string is

$$v_{\text{string}} = \sqrt{\frac{T_S}{\mu}} = \sqrt{\frac{200\,\text{N}}{0.001\,\text{kg/m}}} = 447.2\,\text{m/s}$$

The wavelength of the wave on the string is $\lambda = 2L = 2(0.80\,\text{m}) = 1.60\,\text{m}$. Thus, the frequency of the wave is

$$f = \frac{v_{\text{string}}}{\lambda} = \frac{447.2\,\text{m/s}}{1.60\,\text{m}} = 279.5\,\text{Hz}$$

Finally, the wavelength of the sound wave that reaches your ear is

$$\lambda_{\text{air}} = \frac{v_{\text{sound}}}{f} = \frac{343\,\text{m/s}}{279.5\,\text{Hz}} = 1.2\,\text{m}$$

Assess: This is a reasonable wavelength and the units are correct.

P16.39. Prepare: For a string fixed at both ends, successive resonant frequencies occur at

$$f_m = m f_1 \quad \text{and} \quad f_{(m+1)} = (m+1)f_1$$

Solve: Inserting given information into the previous expressions, obtain

$$325\,\text{Hz} = m f_1 \quad \text{and} \quad 390 = (m+1)f_1$$

Dividing the second expression by the first and for m, obtain $m = 5$.
Knowing that $m = 5$, we can use the first expression to obtain the fundamental frequency.

$$f_1 = 325\,\text{Hz}/m = 325\,\text{Hz}/5 = 65\,\text{Hz}$$

Or using the second expression, we obtain the same value.

$$f_1 = 390\,\text{Hz}/(m+1) = 390\,\text{Hz}/(5+1) = 65\,\text{Hz}$$

Assess: It is good practice to look for simple checks on your work. Calculating the fundamental frequency using both expressions is a simple check.

P16.43. Prepare: For the stretched wire vibrating at its fundamental frequency, the wavelength of the standing wave from Equation 16.1 is $\lambda_1 = 2L$. From Equation 15.2, the wave speed is equal to $\sqrt{T_S/\mu}$, where $\mu = \text{mass/length} = 5.0 \times 10^{-3}\,\text{kg}/0.90\,\text{m} = 5.555 \times 10^{-3}\,\text{kg/m}$. The tension T_S in the wire equals the weight of the sculpture or Mg.

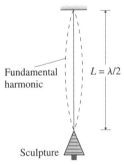

Fundamental harmonic $L = \lambda/2$

Sculpture

Solve: The wave speed on the steel wire is

$$v_{\text{wire}} = f\lambda = f(2L) = (80\,\text{Hz})(2 \times 0.90\,\text{m}) = 144\,\text{m/s}$$

Thus,

$$v_{\text{wire}} = \sqrt{\frac{T_s}{\mu}} = \sqrt{\frac{Mg}{\mu}} \;\Rightarrow\; M = \frac{\mu v_{\text{wire}}^2}{g} = \frac{(5.555 \times 10^{-3}\,\text{kg/m})(144\,\text{m/s})^2}{9.8\,\text{m/s}^2} = 12\,\text{kg}$$

Assess: A mass of 12 kg for the sculpture is reasonable.

P16.47. Prepare This will be a two-step problem. We'll first compute the wave speed from the fundamental relationship for periodic waves, $v = \lambda f$, and then use that in Equation 15.2 ($v = \sqrt{T_s/\mu}$), which we will solve for T_s. We will need to compute μ from $\mu = m/L$, so we will also need to compute m (the mass) from the density and volume of the cable.

Since the cable is vibrating in its fundamental mode and the cable is tied down at both ends, $L = \lambda/2$ for m (the mode) = 1. We are given $L = 14$ m, T (the period) = 0.40 s, and $r = d/2 = 2.5$ cm/2 = 0.0125 m. The density of steel is $\rho_{steel} = 7900$ kg/m^3.

Solve: First, compute μ.

$$\mu = \frac{m}{L} = \frac{\rho V}{L} = \frac{\rho(\pi r^2 L)}{L} = \rho \pi r^2 = (7900 \text{ kg/m}^3)\pi(0.0125 \text{ m})^2 = 3.88 \text{ kg/m}$$

Second, find v from the fundamental relationship for periodic waves.

$$v = \frac{\lambda}{T} = \frac{2L}{T} = \frac{28 \text{ m}}{0.40 \text{ s}} = 70 \text{ m/s}$$

Last, solve Equation 15.2 for T_s and plug in the two previous results.

$$T_s = v^2\mu = (70 \text{ m/s})^2(3.88 \text{ kg/m}) = 19000 \text{ N} = 19 \text{ kN}$$

Assess: The answer is large, but not unexpected for a steel cable holding up a bridge.

Notice in the calculation of μ that L canceled out; this is reasonable because μ should be the same, assuming a uniform cable, regardless of the length.

Note that the units work out in the last equation.

P16.55. Prepare: Please refer to Figure P16.55. The nodes of a standing wave are spaced $\lambda/2$ apart. The wavelength of the mth mode of an open-open tube from Equation 16.6 is $\lambda_m = 2L/m$. Or, equivalently, the length of the tube that generates the mth mode is $L = m(\lambda/2)$. Here λ is the same for all modes because the frequency of the tuning fork is unchanged.

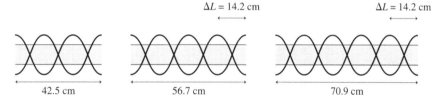

ΔL = 14.2 cm	ΔL = 14.2 cm

42.5 cm   56.7 cm   70.9 cm

Solve: Increasing the length of the tube to go from mode m to mode $m + 1$ requires a length change:

$$\Delta L = (m + 1)(\lambda/2) - m\lambda/2 = \lambda/2$$

That is, lengthening the tube by $\lambda/2$ adds an additional antinode and creates the next standing wave. This is consistent with the idea that the nodes of a standing wave are spaced $\lambda/2$ apart. This tube is first increased by $\Delta L = 56.7$ cm – 4.25 cm = 14.2 cm, then by $\Delta L = 70.9$ cm – 56.7 cm = 14.2 cm. Thus $\lambda/2 = 14.2$ cm and $\lambda = 28.4$ cm = 0.284 m. Therefore, the frequency of the tuning fork, using Equation 15.10, is

$$f = \frac{v}{\lambda} = \frac{343 \text{ m/s}}{0.284 \text{ m}} = 1210 \text{ Hz}$$

Assess: This is a reasonable value for the frequency of a tuning for k in the audible range and the units are correct.

P16.57. Prepare: A stretched wire, which is fixed at both ends, forms a standing wave whose fundamental frequency $f_{1\text{ wire}}$ is the same as the fundamental frequency $f_{1\text{ open-closed}}$ of the open-closed tube. The two frequencies are the same because the oscillations in the wire drive oscillations of the air in the tube.

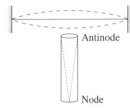

Solve: The fundamental frequency in the wire from Equation 16.5 is

$$f_{1\text{ wire}} = \frac{v_{\text{wire}}}{2L_{\text{wire}}} = \frac{1}{2L_{\text{wire}}} \sqrt{\frac{T_S}{\mu}} = \frac{1}{(1.0\text{ m})} \sqrt{\frac{440\text{ N}}{(0.0010\text{ kg}/0.50\text{ m})}} = 469\text{ Hz}$$

The fundamental frequency in the open-closed tube from Equation 16.7 is

$$f_{1\text{ open-closed}} = 469\text{ Hz} = \frac{v_{\text{air}}}{4L_{\text{tube}}} = \frac{340\text{ m/s}}{4L_{\text{tube}}} \Rightarrow L_{\text{tube}} = \frac{340\text{ m/s}}{4\,(469\text{ Hz})} = 0.181\text{ m} = 18\text{ cm}$$

to two significant figures.

Assess: This is a reasonable length for the open-closed tube. Notice how much a simple sketch can help you understand and solve the problem.

P16.59. Prepare: The waves constructively interfere when speaker 2 is located at 0.75 m and 1.00 m, but not in between. Assume the two speakers are in phase (helpful for visualization, but the result will be generally true as long as the two frequencies are the same). For constructive interference the path length difference must be an integer number of wavelengths, $0.75\text{ m} = n\lambda$, and $1.00\text{ m} = (n+1)\lambda$. Subtracting the two equations gives $\lambda = 0.25\text{ m}$.

Solve:

$$f = \frac{v}{\lambda} = \frac{340\text{ m/s}}{0.25\text{ m}} = 1360\text{ Hz} \approx 1400\text{ Hz}$$

Assess: 1400 Hz is near the "middle" of the range of human hearing, so it is probably right.

P16.63. Prepare: The two radio antennas are sources of in-phase, circular waves. The overlap of these waves causes interference.

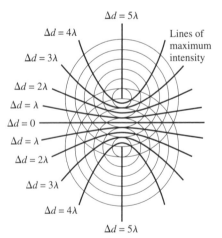

Solve: Maxima occurs along lines such that the path difference to the two antennas is $\Delta d = m\lambda$. The 750 MHz $= 7.50 \times 10^8$ Hz wave has a wavelength $\lambda = c/f = 0.40\text{ m}$. Thus, the antenna spacing $x = 2.0\text{ m}$ is exactly 5λ.

The maximum possible intensity is on the line connecting the antennas, where $\Delta r = x = 5\lambda$. So this is a line of maximum intensity. Similarly, the line that bisects the two antennas is the $\Delta d = 0$ line of maximum intensity. In between, in each of the four quadrants, are four lines of maximum intensity with $\Delta d = \lambda$, 2λ, 3λ, and 4λ. Although we have drawn a fairly accurate picture, you do *not* need to know precisely where these lines are located to know that you *have* to cross them if you walk all the way around the antennas. Thus, you will cross 20 lines where $\Delta d = m\lambda$ and will detect 20 maxima.

Assess: Notice that it would be essentially impossible to solve this problem without a clear and carefully drawn figure. When possible, your solution should include at least a rough sketch. This rough sketch will help you visualize mentally what needs to be done.

P16.65. **Prepare:** The superposition of two slightly different frequencies gives rise to beats.
Solve: **(a)** The third harmonic of note A and the second harmonic of note E are

$$f_{3A} = 3f_{1A} = 3(440 \text{ Hz}) = 1320 \text{ Hz} \quad f_{2E} = 2\,f_{1E} = 2(659 \text{ Hz}) = 1318 \text{ Hz} \Rightarrow f_{3A} - f_{2E} = 1320 \text{ Hz} - 1318 \text{ Hz} = 2 \text{ Hz}$$

The beat frequency between the first harmonics is $f_{1E} - f_{1A} = 659 \text{ Hz} - 440 \text{ Hz} = 219 \text{ Hz}$.

The beat frequency between the second harmonics is $f_{2E} - f_{2A} = 1318 \text{ Hz} - 880 \text{ Hz} = 438 \text{ Hz}$.

The beat frequency between f_{3A} and f_{2E} is 2 Hz. It therefore emerges that the tuner looks for a beat frequency of 2 Hz.

(b) If the beat frequency is 4 Hz, then the second harmonic frequency of the E string is

$$f_{2E} = 1318 \text{ Hz} - 4 \text{ Hz} = 1314 \text{ Hz} \Rightarrow f_{1E} = \frac{1}{2}(1314 \text{ Hz}) = 657 \text{ Hz}$$

Note that the second harmonic frequency of the E string could also be $f_{2E} = 1318 \text{ Hz} + 4 \text{ Hz.} = 1322 \text{ Hz} \Rightarrow f_{1E} = 661 \text{ Hz}$.

This higher frequency can be ruled out because the tuner started with low tension in the E string and we know that

$$v_{\text{string}} = \lambda f = \sqrt{\frac{T}{\mu}} \Rightarrow f \propto \sqrt{T}$$

Assess: It would be impossible to tune a piano without a good understanding of beat frequency and harmonics.

P16.67. **Prepare:** Let's treat this problem as a blinker analog to two speakers emitting sound at different frequencies. Knowing the period of one of the blinkers, we can determine its frequency. Knowing the time interval between the instances when the blinkers are in sync, we can determine the frequency of occurrence of this event, that is, the beat frequency. Knowing the beat frequency, the frequency of the student blinker, and that the other blinker is faster, we can determine the frequency of the other blinker.
Solve: The frequency for the student blinker is $f_s = 1/T_s = 1/(0.85 \text{ s}) = 1.1765 \text{ Hz}$.

The beat frequency is $\Delta f = 1/\Delta t = 1/17 \text{ s} = 0.05882 \text{ Hz}$.

Knowing that the other blinker is faster, it's frequency is $f_0 = \Delta f + f_s = 0.0588 \text{ Hz} + 1.1765 \text{ Hz} = 1.2353 \text{ Hz}$.

The period for the blinker of the other car is $T_0 = 1/f_0 = 1/1.2353 \text{ Hz} = 0.81 \text{ s}$.
Assess: The other blinker is faster so it should have a smaller period. Note that the student blinker will flash 20 times and the other blinker will flash 21 times in the 17 s time interval.

P16.71. **Prepare:** We will need concepts from Chapter 15 to solve this problem. The speed of ultrasound waves in human tissue is given in Table 15.1 as $v = 1540 \text{ m/s}$. The frequency of the reflected wave must be 2.0 MHz $\pm$ 520 Hz, that is, $f_+ = 2,000,520 \text{ Hz}$ and $f_- = 1,999,480 \text{ Hz}$.

There are a couple of mathematical paths we could take, but it is probably easiest to carefully review Example 15.13 and use the mathematical result there, as it is very similar to our problem and gives an expression for the speed of the source v_s.

Solve:

$$v_s = \frac{f_+ - f_-}{f_+ + f_-} v = \frac{2{,}000{,}520 \text{ Hz} - 1{,}999{,}480 \text{ Hz}}{2{,}000{,}520 \text{ Hz} + 1{,}999{,}480 \text{ Hz}} 1540 \text{ m/s} = \frac{1040 \text{ Hz}}{4{,}000{,}000 \text{ Hz}} 1540 \text{ m/s} = 0.40 \text{ m/s}$$

Assess: We may not have a good intuition about how fast heart muscles move, but 0.40 m/s seems neither too fast nor too slow for a maximum speed.

PptIV.21. Reason: It is a logarithmic scale.

$$\beta(\mathrm{dB}) = 10\log\frac{I}{I_0} - 27 \text{ dB} \quad \Rightarrow \quad \frac{I}{I_0} = 10^{2.7} = 500$$

Assess: The answer may be surprising to those who don't realize the decibel scale is logarithmic.